Constructing a Mind

In this book, Professor Imbasciati has built an impressively detailed picture of what he calls the 'protomental' processes, which renders intelligible much that has hitherto been obscure in our understanding of the deepest levels of mentalisation. This is an important book which helps to link psychoanalytic understanding with the wider field of psychology in general.

David M. Black, British Psycho-Analytical Society

Constructing a Mind draws on psychoanalytic theories of mind and recent developments in cognitive science to present the protomental system, a new and original explanatory theory of the development of the human mind.

This book aims to move psychoanalytic theory away from its origins in Freud's theory, toward a model which gives priority to cognition and memory. This, Antonio Imbasciati argues, will make possible a successful and productive integration of psychoanalysis with other areas of psychology. Subjects covered include:

- The mind as an information-processing system
- Constructing the system: from fetus to baby, child, and finally adult
- The caregiver relationship as a decoding system for information processing
- The paranoid-schizoid metabolism of information
- Memory of functions and memory traces of affects
- Internal information generated by the system
- The depressive position and learning to know
- Reparation and thought

This thoughtful and thorough account of cognitive development provides a conceptual framework that succeeds in making some of the more complex areas of psychoanalytic theory more intelligible.

Constructing a Mind will be of great interest to psychoanalysts, psychotherapists and cognitive psychologists, especially those with an interest in neuropsychology and neonatal development.

Antonio Imbasciati is Professor of Clinical Psychology at the Faculty of Medicine and Surgery, University of Brescia. He is a member and training analyst of the Italian Psychoanalytical Society (SPI) and of the International Psychoanalytical Association (IPA). www.imbasciati.it

Constructing a Mind

A new basis for psychoanalytic theory

Antonio Imbasciati

Translated by Philip Slotkin

Routledge
Taylor & Francis Group
LONDON AND NEW YORK

Italian edition entitled *Nascita e costruzione della mente: La teoria del protomentale* © 1998 by UTET Libreria Srl, Turin

This edition published 2006 by Routledge
27 Church Road, Hove, East Sussex BN3 2FA
Simultaneously published in the USA and Canada
by Routledge
711 Third Avenue, New York, NY 10017

Routledge is an imprint of the Taylor & Francis Group, an informa business

Typeset in Times by RefineCatch Ltd, Bungay, Suffolk

Cover design by Lisa Dynan

British Library Cataloguing in Publication Data
A catalogue record for this book is available from the British Library

Library of Congress Cataloging in Publication Data
Imbasciati, Antonio.
[Nascita e costruzione della mente. English]
Constructing a mind: a new basis for psychoanalytic theory / Antonio Imbasciati; translated by Philip Slotkin
p. cm.
Includes bibliographical references and index.

ISBN 1-58391-767-5 (hbk)
1. Psychoanalysis—Philosophy. 2. Cognitive science. I. Title.
BF175.15213 2006
150.19′5—dc22 2005018226

ISBN13: 978-1-58391-767-1 (hbk)
ISBN10: 1-58391-767-5 (hbk)

Contents

List of figures

Note

This translation is based on a thoroughly revised and updated version of the Italian book mentioned on the copyright page and reflects the further development of the author's ideas in the intervening years.

Introduction

The reason for the success of psychoanalysis lay not only in its clinical value (the understanding and treatment of syndromes formerly deemed inexplicable and virtually untreatable), and certainly not in its method, which was thought 'strange' and unscientific, and was criticized as unprofessional and almost immoral. In the scientific world that straddled the nineteenth and twentieth centuries, it was successful mainly because its founder endowed it with a theoretical edifice, the energy-and-drive theory, which offered an explanation of psychic functioning and its disturbances that was acceptable in terms of the sciences of the day – for the concepts of libido, drive, psychic energy, discharge, instinct, economy and the like mirrored contemporary scientific principles and discoveries in the fields of neurophysiology and thermodynamics.

Today, somewhat paradoxically, the method has been accepted and the clinical aspect recognized at least in part, whereas the theory – the energy paradigm – is precisely the element that meets with the most criticism, if not downright rejection. These critiques, voiced even within the psychoanalytic associations, have been expressed for some decades now. Yet Freudian theory seems to be holding up: to this day, the whole of psychoanalysis is identified in the public mind with Freud's drive theory. A possible reason for the persistence of this stereotype, in my opinion, is that, whereas different psychoanalytic models have been successively proposed, no clear *explanatory* alternative to the original Freudian drive theory has ever been proposed. Even today, the energy-and-drive theory retains a heuristic value – that is to say, it serves as a metaphoric model for understanding the dynamics of affects, and is therefore useful for clinical purposes. However, it used also to possess explanatory value, which for Freud may perhaps have outweighed the heuristic aspect, but which is no longer tenable today.

In every science, modern epistemology distinguishes levels of descriptive, interpretative, and explanatory knowledge (Imbasciati 1994). The first two concern, to different degrees, the understanding (*how* and *how well*) of observed phenomena, while the third relates to their explanation (*why*). This last level, to a much greater extent than the first two, requires the explanation to accord with the results of other sciences that deal with the same phenomena from different observational standpoints. In addition to providing a descriptive and interpretative key to the understanding of psychic events (that is, one whereby subjectivity and its development could be understood 'from within'), Freud sought to explain these events 'objectively', in a manner consistent with the discoveries, or, at least, the hypotheses, of other sciences. This was achieved with the energy-and-drive theory, by the reference to instinct, a psychobiological energy, and the laws of its distribution and transformations – that is, of its dynamics and economy. This was consistent with the sciences of the time – in particular, neurophysiology (as in the reflex arc and electrophysiological discharge models) and thermodynamics.

In other words, the explanatory value implicitly attributed by Freud to his theoretical edifice, and probably also the fascination it exerted for many years, rested on its agreement, or consistency, with the neurosciences of the day, so that psychoanalysis seemed effectively to bridge the gap between the human and the natural sciences. Such an agreement no longer exists, for today's neurosciences see the brain not as an apparatus with (bioelectric) energies flowing within it, but as an information-processing system capable of self-modification in a process of development and construction. They also present a picture of the neural system not as a finished, genetically determined apparatus simply waiting to receive and store experience, but as a predisposition to develop certain functions rather than others according to its interactions with the various inputs received. These interactions include the processing of information derived from experiences. The results of this processing determine the constitution and construction of the functions peculiar to a given individual mental system, and proceed in parallel with neural development, which they in fact seem also to determine. In particular, the processing of the primal life experiences conditions not only the functioning but also the very morphology of the neural system, as has been demonstrated experimentally in animals. The vision of the present-day neurosciences is thus one in which both neural and psychic development fundamentally take the form of 'learning processes', especially in the fetus and the neonate; moreover, these processes are not a matter of passive imprinting, but of active processing. Hence the importance of the so-called cognitive sciences and their agreement with both neurophysiology and experimental psychology.

For this reason, Freud's theoretical edifice can no longer be deemed to possess explanatory value, although it can possibly, if understood metaphorically, be retained and used as a model to facilitate clinical comprehension, rather than as a theory proper. In particular, after all, Freud's theory is inconsistent with the findings of the present-day neurosciences. This raises the question of whether it is possible fully to preserve the scientific accomplishments of psychoanalysis – specifically, its method and the relevant discoveries – while at the same time developing a properly psychoanalytic theory that both possesses explanatory value and is consistent with the other sciences of the mind today.

The Freudian theory used the model of a *vis a tergo*, identified as instinct; this was in agreement with the genetic and neurophysiological conceptions of the time, according to which the neural apparatus was seen as developing essentially along biological lines, or at any rate on the basis of organic causes, while experience was deemed to be merely its content. The intimate, mutual connection that has now been established between learning (in the sense of the processing of experiences), consequent acquisitions of mental functions and their encoding in memory, as well as the morphological and functional development of the neural apparatus, was virtually unknown in Freud's day. Neural functioning was seen as relatively independent of that of the psyche, and Freud attempted to unite the two with the concept of the drive cathexis of reality. Drive cathexis was seen as modifying standard neural functioning. Psychic functioning was, furthermore, regarded as dependent on neural functioning, but the converse was not held to be the case. Given this assumption of constant neural function, Freud had to explain the enormous intersubjective variability of the psychic functions he was discovering; he sought to reconcile the unidirectionality mentioned above (neural versus mental functioning) with variability between individuals through the concepts of conflict, suppression, repression, psychic reality as an effect of the drive economy, and other notions included in his theoretical construction.

The idea that psychic functioning is independent of the workings of the neural apparatus was favoured by the prevailing ignorance of the importance of the preverbal psychic functions and of the extent to which their acquisition moulded the corresponding neural functions. This, in my view, is one reason why the separation between affect and cognition, inherited from philosophy, was able to persist for so long: since the prelinguistic mental functions went unrecognized, full consideration was not given to the corresponding neural functions, so that their products in adult psychic life – the affective processes, which already appeared to differ from the more developed mental functions in the subjectivity of the individual – were held to be different also in nature. The result was the distinction, and separation, between cognitive functions, which were regarded as more properly mental, and affective functions, whose nature and origin remained obscure. In other words, the functioning of the nervous system was thought to be predetermined, while the mental functions, fully recognized only after the acquisition of language and labelled 'cognition', were considered to be conditioned by the neural functions and hence to be universal – unless 'disturbed' in a given individual. Such a 'disturbance' was attributed to non-cognitive functions and labelled 'affective'. While interindividual variability was thereby explained, the origin and genesis of the affective functions remained in obscurity. Freud attempted to explain them by the formulation of his energy-and-drive theory.

However, this meant that the affective processes – the salient aspect of all truly mental, albeit prelinguistic, functions – were considered from a fallacious starting point. The reduction of the cognitive processes to those observable after language acquisition and their consequent seeming constancy, at least in the normal situation (today we know how variable they in fact are – cf. cognitive strategies), favoured the view that they were a natural emanation of predetermined neural functions. By contrast, given their conspicuous variability (even within the norm), affects came to be seen as different in nature from the cognitive processes, or, at any rate, secondary to cognition. Furthermore, the mesolimbic brain region was virtually unknown in Freud's day. On discovering the enormous importance of affects, Freud sought to correlate them with all the functions hitherto regarded as mental through the biologistic concept of the drive; affect was seen as the drive's psychical representative as modelled by experience. The concept of drive cathexis explained the extent to which drives and affects are able to influence representations and the cognitive processes; this explanation was necessary at the time because it was otherwise supposed that, as emanations of neural functions held to be immutable, representations and cognitive processes must be constant. So the idea that affect differed in nature from cognition tended to persist in Freud (Imbasciati 1991).

The foregoing follows from the all-embracing nature of the energy-and-drive theory: through the concept of instinct, biology was contrasted with the particular psychic activity that characterizes subjective experience for each individual but was linked to it. Freud's theory was designed precisely with a view to reconciling the great interindividual variability of subjective experience with the universality of biology for all individuals.

Today, the concept of instinct is no longer accepted in its original sense, and is regarded as ambiguous and of little value for explaining the behaviour of living organisms scientifically (Arnold, Eysenck and Meili 1972; Hinde 1974; Imbasciati and Ghilardi 1989; Ghilardi and Imbasciati 1990). Innate predispositions may perhaps be invoked to assign a certain meaning to specific configurations of sensory afferences (as in the classic example of the red spot in a white field for seagulls), and this idea may be associated with the concept of imprinting (Lorenz 1969). These predispositions, however, require learning. In parallel with the abandonment of the concept of instinct, the study of learning processes has developed beyond their representational (let alone verbal) aspect: learning is now understood as any kind of processing of experience that can give rise to some kind of change in the organization governing behaviour. The study of learning is thus extended back to the very beginnings of life, and affects are deemed to be basic cognitive schemata (cf. Plutchik 1980), which are therefore learned.

A consequence of the expansion of the study of learning has been the development of research on the ways in which learning is achieved and on how it is stabilized rather than transformed – that is, the study of cognitive processes. Both involve the study of memory traces and of their 'quality' – that is, of the particular sensory elements of which they are composed – and entail the abandonment of the naive prejudice, still widespread today, that a 'trace' means an 'image' referable to some real object or situation: 'memories' (the use of the plural conveys their affective quality and fleeting aspect) are formless, absurd, disparate, obscure and bizarre. A more detailed investigation is needed of how the various traces, especially the most primitive, are constructed by the processing of experience – that is, by the cognitive processes – how they become stabilized, and, in particular, how each trace impresses its stamp on its successors, in the constant transformations and stratifications undergone by the traces throughout the course of life.

In view of the foregoing, Freudian theory appears inconsistent with the present-day cognitive sciences and with the generality of the neurosciences. The concept of libido, as a *vis a tergo* modelled by the encounter with reality, and Freud's endogenist and innatist model in general, do not sit well together with the accepted empiricist paradigm, with the discoveries of the importance of primary (neonatal and fetal) experience in the construction of the basic mental structures, and with the activity performed from the beginning by these structures on the data of external reality, involving the selection, reading and processing of every item of sensory information. Yet abandonment of the energy-and-drive theory by no means implies, as some fear, a minimization, let alone rejection, of the Freudian edifice, for the method established by Freud and the consequent clinical exploration remain immutable discoveries in the history of the sciences. However, a distinction must be made between method, discoveries and theories (Imbasciati 1994): discoveries remain, theories change and methods are refined – in all sciences. As to theory, Freud could do no more than construct the best possible one in relation to the corpus of scientific knowledge of his time. Theories are neither true nor false, but are hypothetical constructions that may be useful to a greater or lesser extent at a given point in scientific development.

The critiques levelled at Freudian theory in the last few decades, some of them originating within the psychoanalytic scientific community, are therefore more than justified; indeed, they are welcome from the point of view of the possible development of psychoanalysis itself. In my opinion, the studies that seek to compare psychoanalysis with the cognitive sciences are likely to be of particular value. The recent contribution of Bucci (1997), reviewing research in the field of information processing and multiple code theory for the construction and use of memory traces, is extremely relevant to psychoanalysis. However, the author's comparison with psychoanalysis is based on Freud's metapsychology, and disregards the large number of other psychoanalytic theories that have been developed since Freud – in particular, the so-called object-relations theories arrived at by many authors on the basis of Kleinian intuitions. One of the best known, of particular relevance to learning processes, is advanced in Bion's famous work *Learning from Experience* (1962). With their emphasis on the acquisitions of the earliest period of life, the structuring of mental functions in response to the first relational experiences, and the processing of data from the context of the primal sensory experiences, object-relations theories are the ones best suited to comparison of psychoanalysis with the cognitive sciences. For these theories can more readily – and, in my view, necessarily – be considered in relation to memory traces; after all, in present-day neurophysiological terms, the elaboration of relational experience means the processing of memory traces, and hence cognitive processes. The cognitive processes must of course be investigated without recourse to the old reductive notion of a cognition confined to the more developed processes and distinct from affectivity; attention must be directed to neonatal (and fetal) cognitive processes and the mnemic processing carried out on them, which underlies the affective structure. For this reason, in my view, the value of Bucci's work for psychoanalysis, and in particular the reference to multiple-code theory, remains to be developed, precisely in relation to object-relations theories.

Owing partly to the reluctance of psychoanalytic societies to formulate alternative theories to Freud's, a climate of uncertainty, or rather of incomplete definition of theory, has arisen, so that different theories have been deemed equally valid options for every analyst. While this may be useful, and may indeed make for enrichment of the models, it can be an obstacle to the formulation of a concrete theory – as opposed, in my view, to a model (Imbasciati 1994). A theoretical type of formulation may have heuristic value for clinical purposes, or it may help us to describe the observed phenomena, to interpret them, and to transmit them to other workers, and in this sense may also be interchangeable with another, different formulation. However, a true, 'strong' theory must also have explanatory value and as far as possible be unitary. This, I believe, was Freud's intention: to explain the observed phenomena in addition to describing and interpreting them – and to explain them with hypotheses necessarily consistent with the findings of other sciences concerned with the same events, albeit from different standpoints. The Freudian theory was in line with the sciences of the time – hence, to my mind, its success and fascination.

Object-relations theories necessarily emphasize experience rather than development due to endogenous forces; they may therefore assume explanatory value to the extent that the elaboration of experience can be compared with the cognitive sciences' findings expressed in terms of memory traces. Of course, our conception of a memory trace today differs from the old view. A memory trace is now seen no longer as a content, or an 'image' of something, but as the *inscription of a functional acquisition*; the most important traces are those of the functions, which the system learns progressively by the processing of experience. A memory trace has a neural correlate, and we are here concerned with its explanatory aspect, for the manner of establishment of certain neural connections can tell us *why* the corresponding mental functions come into being. However, object-relations theories have been used predominantly to describe and understand *how* the psyche of individuals is structured, whereas the question of *why* has been either set aside or referred – often none too clearly – to the Freudian tradition.

Many authors have developed object-relations theories without reference to the energy-and-drive model, albeit without repudiating it; the entire British school belongs to this tradition. The developments of the Bion school, by emphasizing the aspect of learning *from* experience, imply that the structuring of the mind does not require endogenous forces (such as libido or drives), but takes place through learning operations, whose laws do not necessarily prove to conform to the Freudian paradigm, or to be susceptible of accommodation within the framework of the energy-and-drive theory. Even the concept of aggression, which in the work of Melanie Klein appears to be linked (in my opinion, only formally) to the concept of instinct (the death instinct), is subsequently detached from the instinctual paradigm (cf. Money-Kyrle 1955, 1978), and is replaced in the Bion school by the notion of destructiveness – and this, stripped of explanatory connotations (the question *why*), is used to describe a relational modality (*how*) based on the concept of fantasy rather than that of the drive. All these fundamental twists and turns in the vision of mental processes have not been made properly explicit or adequately connected with the typically cognitive processes.

The developments of object-relations theories have given rise to substantial divergences within psychoanalysis, with the result that many authors have attempted to combine the two models (drive and object), although without succeeding in formulating a genuine 'theory'. Typical examples can be found in the work of Kohut (1971, 1977), Gedo and Goldberg (1973) and Modell (1975a, b). Other authors have explicitly denounced the energy-and-drive hypotheses on which much of the Freudian theoretical edifice rests, rejecting metapsychology in particular; examples can be found in George Klein (1976) and the review by Eagle (1984).

A glance at present-day psychoanalytic theories reveals so great a many-sidedness and diversity of models, concepts and terms that one may suspect the existence not of 'one', but of 'many' different psychoanalyses (Wallerstein 1988). There has been much questioning of what is meant by psychoanalysis: is it a theory? a technique? a hermeneutics, or indeed an art? Or is it a science? If the last, it ought to be characterized by the homogeneity and consistency of its specific method. I believe that the common ground referred to in a different context by Wallerstein (1990) is to be found precisely in the method. But how is that method to be defined vis-à-vis the – often highly involved – overall structure of this very particular science? Too often, a clear distinction is not drawn between method, techniques, models and theories.

There is a kind of glue that seems to hold the complex and variegated Freudian corpus together in the name of science, and that in the past constituted perhaps the most significant aspect of the scientific comparison of psychoanalysis with other sciences. That glue is the importance Freud assigned to his libido theory, with its associated metapsychology, and in particular to the energy-and-drive conception of the psyche, its origin and development. This 'general theory of the mind' offered not only a description but also an explanation of psychic processes, and thereby placed psychoanalysis on the same level as the sciences of the day. Freud's energy-and-drive theory thus 'held psychoanalysis together' for decades, at first scientifically (because it was syntonic with contemporary scientific theories), and later charismatically.

Between the 1930s and 1950s, the divergences were, so to speak, camouflaged (Greenberg and Mitchell 1983). By the 1960s, the 'witch metapsychology' came to be radically criticized (Fabozzi and Ortu 1996), and was rejected by many, particularly in the USA, following the work of George Klein, Peterfreund, Gill, Rubinstein and others. The attempts to reconcile a range of widely varying conceptions (see Kohut) were followed by alternative suggestions; the information-processing paradigm of Peterfreund (1971) presents a certain interest owing to its pioneering aspect. Freud's descriptions of infantile development were criticized as adultomorphic and pathomorphic (Fossi 1984), and those of the internal world as anthropomorphic. Yet drive theory continued to be upheld by many analysts for a few decades more, and is perhaps still accepted today, even if for doctrinaire rather than scientific reasons – or, better, affective reasons within the psychoanalytic community.

To a clinician, on the other hand, the energy-and-drive theory, with the concept of instinct, appears to be the most logical and natural model in existence, which ought therefore to be espoused, because it is homologous with an adult's conception of the affects and emotions by which he[1] feels governed. After all, a mature adult who achieves the capacity to experience and feel – that is, to represent to himself – his own affects as a dynamic of inner trends will tend to think that the same occurs in all individuals, and that this dynamic does not depend on the subject's own capacity for representation, but is inherent in the nature of affects. The idea that affects have a dynamic helps us to understand a dynamic that we presume another may in turn become capable of feeling. That is the traditional supposition that underlies our interpretations. But it certainly does not mean that affects *really are made up* of these 'forces', which an adult experiences – that is, can represent to himself[2] – inside himself. I have written elsewhere (Imbasciati 1991)

1 [Translator's note: For the sake of brevity and to avoid clumsy constructions, where applicable the masculine personal and reflexive pronouns and possessive adjective are used throughout this book for both sexes.]

2 'Representation' is, of course, meant here not in a conscious sense, still less as a name for conscious affects, but as the possibility, which adults usually have, of gaining insight, often with the help of analysis, into their own internal world.

of the hypostasis of lived experience as an arbitrary element in Freudian theorization.

Taking the energy-and-drive theory on its explanatory level (Freud hoped that the biochemical substrate of the drives might one day be discovered – cf. Freud 1895d [1893–95]: 200ff.; 1905d: 167, 215 and 217; 1905e [1901]: 113; 1906a [1905]: 277; 1914c: 78; 1915c: 125; 1916–17a [1915–17]: 321), we are led to believe that there *are* instinctual 'forces', which *really do* act as an adult experiences them – that is, as he has become capable of representing them to himself. However, the possible form of one's experience of affects does not necessarily imply a homologous type of functioning: the capacity to represent affects to oneself is not the same thing as the existence and functioning of affects. The subjective experience of affects depends on a representational capacity that is a more mature function than the establishment of the functioning of affects as such; it is the capacity to have the experience of the self, the acquisition of subjectivity (Ogden 1990), and a function subsequent to the basic function of the affects, which is acquired, so to speak, as a meta-affective function of affect, during the course of the developmental cycle of the older child and the adult. Small children do not yet possess this function, and many of our adult patients also lack it (Greenspan 1997). We often make the mistake of projecting into these patients our own way of representing affects to ourselves; failing to realize that this is an acquisition of normal, mature persons, we take it for granted that this capacity is inherent in the nature of all human beings' affects – and so we think that all patients, once their defences have been demolished, will once again feel affects in this way. So it is, as Greenspan (1997) points out, that many patients, while seemingly accepting our interpretations, remain untouched by them because they have *never acquired* any way of representing affects to themselves. Yet affects not only exist in these patients, but also modulate their behaviour. The problem is that affects *are* not natural 'forces', even if a mature adult may represent them to himself as such. They cannot therefore readily be linked, on the explanatory level, to the concept of the drive, even if, because they are learned as basic behavioural schemata (Plutchik 1980), they will determine all subsequent learning and hence govern the individual's behaviour.

In my opinion, psychoanalysis has been harmed by its retention of the Freudian theoretical edifice, with its appearance of explanation. That edifice has been hedged with numerous detailed reservations to allow it to be preserved at all costs (Greenberg and Mitchell 1983). This has made the problem of the scientific status of psychoanalysis appear as though it were inherent in the theory, so concealing the discipline's true scientific asset – namely, its method. Moreover, this ambiguity goes hand in hand with a willingness on the part of analysts, in the name of a plurality of techniques, to put up with an inappropriate distinction and definition of terms and concepts – including, first and foremost, the concept of an actual psychoanalytic 'theory'.

Over the many years since I began work in this field in 1978 (Imbasciati and Calorio 1981), I have become increasingly convinced that a 'strong' theory with potential explanatory value, consistent with the neurosciences and with the cognitive sciences in particular, can be derived from the neo-Kleinian matrix of object-relations theories. The need, in my view, is to determine the manner of composition, or rather construction, from a variety of disparate inputs that must also be identified, of the memory traces corresponding to the psychoanalytic notions of the primary relationship with the mother, the internal object, the 'breast', differentiation of the early self, symbolization, and so on. The 'learning *from* experience' intuited by Bion must be translated into terms of memory traces.

The theory presented in this book, in which the mind and its development are considered in terms of learning operations and memory traces, has an explanatory aim; it is intended as a possible successor to Freud's original *explanation*, and to be consistent with the psychoanalytic *descriptions* given, in particular, by more recent authors, as well as with the findings of the present-day neurosciences and cognitive sciences.

Milan, March 2004
Antonio Imbasciati
www.imbasciati.it

Chapter 1

The mind

1.1 Affect, cognition and consciousness

'Mind' is defined differently according to one's standpoint, which may be that of philosophical research (Money-Kyrle 1956, 1961; Meltzer 1978), neurophysiology (Hebb 1949; Benedetti 1969; Eccles 1970; Luria 1970, 1974, 1976; Brown, J. 1977; Boddy 1978; Mancia 1980) or the psychological sciences. For psychologists, as a first approximation, the epithet 'mental' can evidently be applied not only to thought and memory, but also to the complex of feelings and emotions that substantially influence the cognitive processes, even though these processes are seemingly the most elaborate and creative aspect of the mind. The fact that affective-type processes can 'influence' cognitive ones is universally acknowledged, even by laypersons, although there is less agreement on how and with what results. The complex of feelings, affects and emotions is often regarded as a set of psychic processes different in nature from those of cognition (thought, memory, intelligence, etc.), so that the use of the term 'mental' tends to be restricted to the latter, although it is accepted that the former, too, are psychological; in other words, since they take place in the mind, it is conceded that they must be deemed mental. The different nomenclature is reflected in the common usage of the two terms 'psyche' and 'mind'. The former is the more comprehensive, covering all psychological processes, whereas the latter tends to denote certain psychic processes only – namely, the cognitive ones.

Inherent in the different names is a difference in the connotations of the two types of processes. The term 'mental' connotes a 'higher' level and confers privileged status on the processes to which it is applied, so that these tend to be seen as independent of other psychic phenomena – which, for their part, are placed on a lower, secondary level in the mental economy, subordinated to the 'higher' functions. However, this subordination is contradicted both by experience and by common sense, which is perfectly prepared to acknowledge that feelings, affects and emotions often constrain the so-called higher mental processes, and may even actually inspire or govern them. Yet even if the interdependence of the two kinds of psychological phenomena is granted, they are often still contrasted with each other – as, for example, the phrase 'to be overwhelmed by passion' indicates. Here the affective processes (passion) – seen as 'lower' than the processes that 'ought' to be in charge of the human mind and behaviour – are felt to become so much stronger than these 'higher' processes that they overwhelm them; the antithesis between affect and reason is obvious. This opposition is reflected in usage, where the term 'mental' is often felt to be inapplicable to the affective processes.

This view of the mind as divided into two parts stems from the medieval conception of psychology, inherited from the Greek philosophical tradition, as the science of the soul in which human activity is governed by the two contrasting principles of matter and spirit (Hutten 1962); this division has persisted in the lay mind and also, to some extent, in psychology. Although the development of the psychological sciences has amply demonstrated the importance of affective and emotional processes and the close interdependence between them and the cognitive and conative processes,[1] the division in a sense lives on and is expressed by the different connotations of the terms 'mental' and 'psychic'. Another reason for this division is that, from an initial, superficial examination of our subjectivity, cognition appears to possess the character of consciousness, whereas affects (as well as the meaning of actions) more often prove to be concealed. In our view, this division is unwarranted. The simplest and most obvious example is afforded by infant observation: can cognitive activity be distinguished from affective behaviour, or the expression of affect, in a 1-year-old child? Or in a 6-month-old baby? There is no doubt that the infant (*infans* – Latin for 'unable to speak') has cognitive activity: he is getting to know the world! He is learning many things, images, perceptual and recognition capacities (that is, memory), functional skills and the ability to communicate; and he is learning to engage in dialogue with his caregivers, first by eye contact, then by motor activity, next by lallation, and finally in words. Variations in cognitive functions between individual infants are obvious: the expert eye can quickly tell a retarded from a precocious child.

Yet the mental activity of small children is customarily regarded as 'affective' rather than cognitive. A continuum in fact exists between what appears as affect in young children and what appears as cognition in adults: one need only consider the course of development of an individual from the age of 1 year on, through infancy and the later stages of childhood, as he gradually acquires 'rational' capacities and comes to know what he is thinking – that is, as he acquires consciousness. I say 'appears' because the distinction between affect and cognition is only a seeming one, and because the difference

1 The neurosciences too have established that the so-called affectivity centres in the brain modify the cognitive functions; for example, the amygdala regulates memory.

between the two orders of mental processes lies not in their nature, but in the different appearances assumed by affects and rational cognitive processes in adult consciousness. As stated in the introduction, the legacy of the distinction between affect and cognition in our culture is that, once we are adults, some of our mental processes appear to be clear – that is, perfectly conscious – whereas others seem veiled or obscure. In other words, the distinction is based on differences in the degree of consciousness. However, the different appearances in adult consciousness by no means signify that the two types of mental processes are *really* different. So-called affects have to do with cognitive functions, as can plainly be seen in infants and children. According to authoritative studies (Plutchik 1980), affect can be defined as a basic cognitive schema. Kellerman (1980) uses the term 'basic cognitive category'. Again, in the investigation of non-conscious mental processes, not only psychoanalysis but also the 'cognitive sciences' have for decades been aware of a continuum between seemingly affective processes and others that appear cognitive; while obvious in the course of individual human development, this continuum can also be demonstrated in adult mental activity.

The epistemological and methodological problems raised by these distinctions and the relevant research are extremely complex. A 'cognitive psychology' has now existed for some decades, but the relevant issues are beyond the scope of this book: interested readers should consult, for example, Neisser (1967, 1976); Kanizsa (1975); Hamilton and Vernon (1976); or Kanizsa and Legrenzi (1978). In our own approach, as will become clear in the course of this book, all psychic processes are considered from a single viewpoint combining cognitive and psychoanalytic aspects, within the framework of the theory to be outlined below (see Chapter 2). For this reason, the term 'mental' will be used to denote all psychological processes, including those defined by others as affective and/or emotional.

As will be seen, the term 'mental' – denoting the characteristic element of psychic processes – is here applied not only to conscious manifestations but also to a whole range of phenomena that do not attain the level of consciousness. From the very first psychoanalytic discoveries at the end of the nineteenth century to this day, the development of the psychological sciences has increasingly revealed the existence of non-conscious mental phenomena. Unconscious processes are in fact much more numerous and perhaps more important than their conscious counterparts. For this reason, consciousness is regarded as an attribute of certain mental processes, and accessibility to verbalization as only one aspect of the activity of the mind, not encompassing the whole of psychic reality (Lunzer 1979; Davidson and Davidson 1980). Any conscious mental event presupposes unconscious processes, so that conscious phenomena are seen as the terminal event in a chain of unconscious processes; there can be no conscious thought unless there has first been an unconscious thought (Bion 1962). This statement, which is manifestly true on the diachronic level of individual development from birth to adulthood, is assumed here also to be valid synchronically for any adult mental event. On this point, we agree with Bion (1962, 1963, 1965) that the genesis of any activity definable as thought predominantly involves unconscious processes: every thought presupposes and contains within itself a chain of unconscious processes, from the most primitive to the most highly developed, which may, but do not necessarily, assume the quality of consciousness (as a mere epiphenomenon). That is the sense in which Bion's thesis should be understood.

Other authors hold that consciousness is only one of the possible manifestations of neurological functions, albeit not a constant element in their activity (Brown, J. 1977; Arieti 1969; Davidson and Davidson 1980). The mental character of a neurological process is unequivocally betrayed by the quality of consciousness; however, consciousness does not introduce the 'psychic' element, although it incontrovertibly demonstrates it. Yet this is demonstrable, if less readily, at an even earlier stage – on the level of processes that cannot be regarded as mental because they give rise, for example, to complex behaviour patterns that, while seemingly automatic, indicate the presence of ordered, coordinated processes of even greater complexity, albeit ones that are not conscious. Consciousness is therefore not an intrinsic property of neurological functions, but just one of the many mental functions acquired and performed by a neurological system.

The distinction between affect and cognition has also been upheld by psychoanalysis, especially in the Freudian theory,[2] in so far as their seeming differentiation in adult consciousness has taken on a substantial quality (hypostasis of experience – cf. Imbasciati 1994) in the energy-and-drive theory: here affectivity, although studied in terms of the continuum between it and cognition, is linked directly to a psychobiology of instincts and deemed indirectly to influence the cognitive processes, which are seen as a natural consequence of brain function. Hence the particular conception of the unconscious maintained by Freud and the early psychoanalysts. On discovering unconscious processes, Freud asked, 'Why the unconscious?', and, as we shall see, based his theory on this question. Such a question in fact assumes a priori that what is mental must be conscious; from our present-day vantage point, therefore, the wrong question has been asked. Today, we should put it in these terms: 'Why consciousness?' In other words, why, in the development of its mental functions, does the human species also acquire the highly particular self-reflective function that is called consciousness? This reversal of viewpoint on a fundamental question has given rise to still continuing radical changes in the development of psychoanalysis.

2 Psychoanalysis is by no means synonymous with the Freudian theory, which is merely the first to have been formulated in the development of psychoanalytic science.

The term 'mental', then, covers not only the events traditionally and restrictively defined as such (e.g. intelligence), but also emotional and affective events; it is extended, too, to non-conscious processes. It is applied, in addition, to a range of phenomena that are not ordinarily seen as mental – that is, to all forms of human behaviour and action. While this seems obvious for complex patterns of behaviour, especially those governing interpersonal relationships, simpler behaviours too involve a mental element. Even riding a bicycle from A to B can be regarded as a mental phenomenon, as can someone's behaviour in punching, hugging or having sexual intercourse with another person. All these behaviours can be defined as mental events in so far as they are performed in accordance with a specific, consistent program (either automatic or established on the spot, but in any case always acquired), which cannot be regarded as anything but mental, whether or not it is rational and conscious in nature. What seems to be the mark of a mental function is a *program*.

It must not be forgotten that the mind is not a 'substance' (the animic substance of medieval philosophy), but the set of functions that control every activity and manifestation of the individual. This set is acquired, as will be shown in detail later, by progressive learning operations. It is supported by the neural functions, but is not their direct, obligatory emanation for all individuals: the neural functions are modulated by learning operations, and it is these that record the successively acquired functions in the neural apparatus. Even the morphology of the brain is determined by learning operations. For this reason, a program, as the key concept in the definition of a mental function, must correspond to the notion of a memory trace (a 'trace' refers to the acquisition of a given function). This correspondence between the concept of a functional program (which is acquired) and that of a trace recorded in the neural structure is the foundation of our theory, which infers the corresponding construction of successive functional traces from the psychic processes investigated by the psychoanalytic method. Our aim is to identify a psychophysiological counterpart to the findings of psychoanalysis.

Even much simpler manifestations than those usually described as 'behaviour patterns' or 'actions' – for instance, any motor activity – must be regarded as mental. Someone who takes out a cigarette and lights it, even if absent-mindedly, performs a series of movements that depend on a precise muscular sequence, set in train by a program contained in his mind – that is, by a set of processes that may or may not have become automatic but are in fact the result of a precise learning operation and can be modified at any time by other mental processes. In walking, too, one performs a series of actions which, while automatic, share the mental properties of the previous example. Considered in these terms, even the simplest motor action, such as bending one's forearm from the elbow, cannot strictly speaking *not* be regarded as a mental event. After all, to bend one's arm, *some* muscle fibres of *some* muscles must be contracted, and *some* must have a certain degree of contraction while *others* must contract to a different degree; at the same time, *other* fibres, belonging to *other* muscles, must – not, of course, arbitrarily – relax, and their reduction in tone must match the increase occurring in the contracted muscle groups. Moreover, all this must be appropriate to the specific type of bending, with a *specific* intensity, a *specific* and precise force, a *specific* velocity and a *specific* amplitude, *specifically* stopping at a *specific* point, with a *specific* and precise deceleration at the end and a *specific* and precise acceleration at the beginning. In other words, an entire, highly complex muscular sequence must be implemented – and an infinite variety of sequences can give rise to an infinity of bending operations with an infinity of different characters. Such sequences are set in train by equally precise sequences of nerve impulses generated by a precise program; in psychological terms, such events correspond to processes definable as mental, some of which may appear automatic and others voluntary,[3] some conscious and others unconscious, some investigable by methods of a certain kind (e.g. the psychoanalytic method, among many others) and others by other methods, whereas still others are relatively inaccessible.

A person's consciousness is therefore a secondary, relatively non-defining, characteristic of the mental; moreover, it is subjective and as such extremely variable. The unconscious of psychoanalysis is a state of lack of consciousness of a very particular kind – that is, it is defined by the specific method used to observe it. Conversely, what appears to be a more 'objective' and general characteristic of the mental is the non-chance nature of many behaviours and behaviour patterns, from which it is possible to demonstrate, or infer, the operation of programs governing the observed manifestation in interaction with the environment.

1.2 Perceptual activity

The term 'mental' can therefore be extended to all behavioural events, from the simplest motor phenomena to the most complex behaviour patterns; the common denominator of mental function appears to be the fact that it is underlain by a 'program' – which indicates learning and memory. However, there is also another class of phenomena that can be regarded as mental in these terms. So far, our analysis has been confined to events that presupposed some kind of mental activity, which was therefore manifested more or

3 If the above example is considered in neurological terms, voluntary – pyramidal – activity is distinguished from extrapyramidal automatism and tone. The argument as to whether extrapyramidal automatism can be regarded as mental is more complex, involving as it does the respective extents to which extrapyramidal motricity is acquired as an exclusive consequence of neural maturation, on the one hand, and as a result of the exercise of functions in the fetus and the neonate, on the other hand.

less plainly on the outside; in other words, we have concentrated, as it were, on the 'output' dimension, from the mind to the outside world. Yet there are also mental events on the 'input' side, such as the phenomena of perception, to which the adjective 'mental' is also applicable.

Perceptual phenomena are 'mental' events in their own right, even if, owing to their simple and seemingly automatic appearance, they are commonly regarded more as physiological events than as genuine mental processes, or rather 'activities'. It is often assumed that perceptions are the automatic result of the functioning of the sensory receptors, acting, in effect, as faithful reproducers of external reality, which is deemed to be 'taken in' by the sensory apparatus and transferred to the mind, where it is used for processes of knowledge. This view, described as 'naive realism' by Gestalt psychology (which demonstrated that it was untenable – Koffka 1935; Metzger 1954; Kanizsa 1975), rests on the basic assumption known as the 'constancy hypothesis' (Katz 1951), according to which objects are perceived as they are in objective, physical reality, so that the subjective experience is characterized as a 'copy' of reality. This is therefore deemed to be the mode of functioning of normal perceptual processes, whereas all other cases, where the percept does not perfectly correspond to the objective reality, are seen as exceptions, or anomalies of perception, the paradigm cases of which are optical-geometrical illusions (Girotti and Calorio 1974; Calorio and Purghé 1977; Purghé and Imbasciati 1979). This theoretical approach is the legacy of a philosophical tradition in which the external world is held to present itself to us as it actually is, and in which knowledge proper consists not so much in the recording of what is seen or heard or touched – which is deemed 'natural' – as in the subsequent use of the perceptual events.

Experimental studies of perception have led to different theoretical conceptions (Allport 1955; Brunswik 1956; Piaget 1969; Carterette and Friedman 1974–8), in which perception is no longer regarded as a direct representation of the 'truth', but instead as a complex process of restructuring and integration of sensory information; the emphasis is laid on the involvement of functional factors that depend on the contextual dynamics of the perceptual act. So-called correct perception raises the same problems as its incorrect counterpart: there is never an exact, faithful reproduction of the stimulation situation, and the seemingly automatic nature of the process is due not to deterministic functioning of the receiving apparatus but to the speed at which the data from the receptors are used in mental processes, the result of which is the subjective event called perception. The central element of perception is a 'recognition' of the object, or of its parts or qualities, which is correlated in a complex manner with the processes of learning and memory. Eccles (1970) summarizes the problem of the constitution of perceptual experience as follows:

> In response to sensory stimulation, I experience a private perceptual world which must be regarded, neurophysiologically, as an interpretation

> [a psychological interpretation – *AI*] of specific events in my brain. Hence I am confronted by the problem: how can these diverse cerebral patterns of activity give me valid pictures of the external world? Usually this problem is discussed in relation to visual perception [. . .]. There seems to be an extraordinary problem in explaining how information from my retinae when relayed to, and activating, my cerebral cortex gives me a picture of the external world with all its various objects in three-dimensional array and endowed with brightness and colour. This epistemological problem has led to much philosophical confusion when it has been discussed on the assumption that fully patterned visual perception is an inborn property of the nervous system. On the contrary, my visual perception is an interpretation of retinal data that in a lifetime of experience I have learned to accomplish, both particularly in association with the sensory information provided by receptors in muscles, joints, skin and the vestibular apparatus, and with the central experiences of willed effort. (Eccles 1970: 66)

And, we may add, this interpretation is also associated with the experience represented by memory.

The problems of perceptual processes have always constituted a wide-ranging field of study, and indeed a discipline in their own right (the psychology of perception, or perceptology). This is therefore not the place for a discussion of its principal aspects, of the complex issue of empiricism versus innatism – the relations between perception and learning – or indeed of the epistemological problems of the process of knowledge in general. Instead, let us merely mention here that the concept of 'reception' can usefully be distinguished from that of 'perception'. The former term usually denotes the physiological process whereby a receiving apparatus, by virtue of its specific neurophysiology, receives certain signals from external objects and conveys them via neurological pathways to more central processing stations, where they undergo more complex processes, of which we may observe the subjective aspect – perceptual phenomena – and which can be described in psychological rather than physical terms. In the usual sense of the word and in accordance with its etymology (Latin *per* + *capio*), a perceptual event possesses the characteristic of consciousness; however, this is not the only subjective event involved in the psychic – that is, mental – processing of the received sense data. The existence of mental events associated with the reception of signals from the external world but lacking the character of consciousness can also be demonstrated; such phenomena range from so-called subception (wrongly described as subliminal perception) to intermediate phenomena in which the characteristic of consciousness is attenuated – for example, when one remembers having seen an object without having registered it at the time of observation (the 'attention/perception' problem).

Because a sense datum can be used for psychic – mental – processes without the objective characteristic of consciousness necessarily being involved, perceptual events, like any other mental processes, can be considered in the context of psychoanalysis – and, in my view, more precisely, within that of Bion's theory. Perception can be regarded as a kind of conscious thought which, like all such, is the – albeit not always necessary – end result of a chain of unconscious processes (Ehrenzweig 1967, 1975). These may account for the entire range of perceptual phenomena involving a discrepancy between the percept and the external object, or, rather, between the percept and the set of stimuli presented by that object to the receiving apparatus. Such phenomena range from the perceptual events studied by Gestalt psychology, via optical-geometrical illusions and the processes involved in projective tests, to the projection phenomena studied by psychoanalysis, and may be grouped together under the heading of apperceptive distortion (Imbasciati 1967, 1978a). The discrepancy that always exists between percept and objective reality gives rise to considerable epistemological problems; in psychoanalytic terms, it shows that perception, as usually understood, is the final event in a sequence of unconscious mental activity, while, on the other hand, the study of the unconscious processes underlying perception can supply information on the reasons for the processes of apperceptive distortion and on their characteristics.

Perceptual activity centres on the process of selection, organization and filtering of information performed by the various neuronal stations along the nerve impulses' pathway from the receiving apparatus to the brain, where they are transformed by certain precise programs into, for example, 'visual thinking' (Arnheim 1970). This therefore entails not the mere mechanical recording of data, but the 'understanding' of meaning structures through a process of active exploration: 'Looking at the world [proves] to require an interplay between properties supplied by the object and the nature of the observing subject' (Arnheim 1974: 6); and 'All perceiving is also thinking, all reasoning is also intuition, all observation is also invention' (ibid.: 5). In other words, perception is a *reading* process, which presupposes the existence of reading units and reading programs – in turn involving learning and memory. If this is so for the perception of objects, whose laws of formal organization have been studied by Gestalt psychology, it is all the more true of more complex situations, such as interpersonal perception (Warr and Knapper 1968). Consequently, for example, when we meet someone unknown to us, of whom we obtain only a vague impression, perhaps limited to an overall black-and-white judgement (pleasant/unpleasant or likeable/not likeable), our reaction results from the complex operations of selection and organization to which we subject the signals from the person concerned. The diversity of impressions depends on the functioning and construction of our personal processing structures (Money-Kyrle 1961; Abercrombie 1969; Eiser and Stroebe 1972), which determine how we organize and perceive the 'silent messages' (Mehrabian 1971) emanating from the unknown person, such as tone of voice, facial expression, the memories he arouses in us, his way of moving and dressing, and so on.

Considered in these terms, then, perceptual activities too can be assigned to the 'mental' domain, irrespective of their degree of consciousness and complexity. They entail precise programs for the selection and processing of information, which not only adapt and recognize external reality, but also inevitably distort it. A 'reading' is never the faithful transposition of a reality presumed to be objective, but the execution of a program for interpretation of the input.

1.3 Instinct or learning?

It is clear from analysis of the above examples that apparently simple phenomena, such as those inherent in perception and motility, are not by any means automatic, but stem from a complex structure involving learning, and hence mental processing. On the other hand, there are also highly complex, organized and goal-oriented behaviours with a substantial automatic element, which cannot be influenced by individual experience. These are the behaviours we call instinctive.

Ants, for example, display highly organized behaviour with functional differentiation, including programmed sequences for food gathering and storage, nest construction, and care and feeding of larvae. Because this complex organization is a characteristic not only of the group but also of the individual, ants have been regarded as 'intelligent'. Yet these phenomena cannot legitimately be assigned to the field of mental activity, because they are not susceptible to modification on the basis of prior experience. Experiments have shown that an ant systematically deprived of the fruit of its labours continues to work until it dies, with an obligatory, programmed sequence of actions that cannot take any account of what has happened.

Similarly, the activity of bees – which possess a sophisticated language (von Frisch 1938 [1933]) capable of precisely indicating the distance and direction of flowers with ripe pollen – suggests the existence of a kind of 'intelligence'; however, this proves not to be susceptible to the influence of experience, but exhibits instead the immutability of repetition. Again, if a hen's eggs are replaced by round, white stones, it will continue to brood them although it possesses instruments capable of distinguishing between an egg and a stone. A vast number of other examples from the animal kingdom present characteristics of complexity and unmodifiability that can be defined as intelligent only in the sense of an organization innate to the species, and hence in the sphere of biological functions.

For instance, one species of wasp feeds its young by catching caterpillars, which it paralyses by injecting a small dose of venom into their ganglia; the paralysed caterpillar, which, however, is still alive, becomes food for the larvae that will hatch from eggs laid by the wasp beside it. While this behaviour is highly complex and organized, it is neither learned nor modifiable: the wasp seems to know exactly what type of caterpillar to catch and which are the crucial parts of its body to attack – but this 'knowledge' has never been 'learned'.

Sexual behaviour is very often adduced to exemplify instinct; in common parlance, the term 'instinct' more often than not refers to sexuality. Yet there are major differences in the role of instinct in sexual behaviour according to position on the zoological scale. Such behaviour in lower animals is found to be rigidly programmed, identical for all individuals of a given species and unmodifiable by learning. For instance, many species of insects and spiders have characteristic sexual 'dances' – that is, fixed, ritual movements that precede mating and betray the obligatory, immutable character of instinctive reproductive behaviour: in many species, mating is followed by the death of the male, who is often devoured by the female, and the individual cannot correlate the event of 'death' with the instinctive behaviour of mating, or modify it. In fish, too, there are interesting examples of the rigid programming of sexual behaviour in a manner that cannot be influenced by learning. The male three-spined stickleback, a freshwater species, assumes a different coloration from the female at spawning time, thus triggering a series of interactive behaviours between the two sexes, which previously seemed alien to each other. Male and female now take turns at nest-building, after which the female lays her eggs in the nest; the male then fertilizes them, and the female (who would otherwise eat them) looks after them until the fry hatch and are able to depart. Among birds, too, there are examples of complex behavioural sequences involving courtship, mating, nest-building and the raising of young.

However, what is actually instinctive proves to be not sexual behaviour itself but the sexual behaviours of certain animal species. Ascending the zoological scale, we find that sexual behaviour can be modified by experience, and not only in its secondary aspects, but sometimes also in the very possibility of performing sexual functions. The monkeys studied by Harlow and Harlow (Harlow 1961; Harlow and Harlow 1965) are a telling example. These authors raised groups of monkeys in different experimental situations, and the results can be summarized as follows. One group were raised normally, the baby monkeys being kept with their mothers, with normal suckling, delousing and care; a second group were brought to the mothers only for suckling; a third were kept in a cage with an artificial cloth monkey and provided with feeding bottles on which the babies could suck; while the fourth and last group were raised in a cage containing a wire frame with feeding bottles. The monkeys in the various experimental groups, when fully grown, were found to exhibit different sexual behaviours: only the normally raised monkeys showed the typical behaviour of their species, while all the others presented difficulties proportionate to the degree of difference from the normal raising situation. Sexual behaviour was thus progressively disturbed, and ultimately prevented altogether or replaced by aggressive behaviour that precluded mating.

It may therefore be concluded that the different experiences of the babies altered the organization of a behaviour – mating – that ought by definition to have been instinctive, automatic and identical for all species. The diversity of prior experience is the norm for man, in whom the behaviour most commonly said to be instinctive – sexual behaviour – is in reality learned (Imbasciati 1998), its individual variability depending on personality, cultural level, cultural background and personal experience. The Harlows' experiments show that sexual behaviour in higher animals is learned: maternal care, particularly body contact, allows the formation of very primitive internal structures that determine adult sexual behaviour. This is demonstrated even more clearly by the clinical treatment of human sexual pathology.

According to a prejudice still widespread today, a mental event – often a motor or perceptual event, or sometimes an emotional one – is said to be instinctive merely because it appears automatic and simple to the subject. In reality, the subject cannot have any awareness of the complex processes that have given rise to events that he sees as simple – or of the learning processes that have made the phenomenon in question automatic. Such learning processes are not only unconscious but often also date back to the earliest stages of development. Not even the fact that a mental event is observed in all individuals of a given species suffices for it to be deemed instinctive – that is, genetically determined. It is quite possible that all individuals of a given species have the same experiences – for instance, tactile experiences (as with the Harlows' monkeys) – which trigger the same early learning processes in them, these processes subsequently being used automatically.

What makes a behaviour instinctive – that is, biologically determined rather than processed at the psychic level so that it is the result of a 'mental' operation – is therefore not the fact that it belongs to a sphere defined a priori as instinctive; for the concept of instinct in higher animals can easily mislead (Imbasciati and Ghilardi 1989; Ghilardi and Imbasciati 1990). Nor is it the simplicity or complexity of a behaviour that determines whether it should be regarded as learned and intelligent on the one hand or instinctive on the other hand; and other irrelevant factors are goal orientation or seeming automatism. Instinctive behaviour presupposes neurological, chromosomally transmitted, automatisms that are immutable and fixed for all individuals of a given species. Such automatisms may relate even to highly complex behavioural sequences, which nevertheless retain the character of a neurological reflex, and therefore remain in the biological field – even if their specific goal orientation may, in some species, cause them to appear intelligent. The mark of the mental, on the other hand, which distinguishes it from the biological, is *learning*, defined not only in the more restrictive and traditional rational sense, but also as the acquisition of primary functional structures, as demonstrated by the Harlows' experiments.

These structures, which are functional, acquired and not biologically determined, can be traced back to the affective and emotional organization, whose extreme variability between individuals clearly reflects its origins in individual experience. However, when the concepts of learning and the associated memory are extended back in this way to such primitive events, it proves difficult to determine their origin. In my view, as I shall attempt to show in more detail later, these structures correspond to the primitive internal objects of psychoanalysis. If the concept of learning is broadened in this way, it is very hard to establish how much of each relevant individual event is truly learned and how much is transmitted genetically. Even biological structure, always present as the substrate of mental life, may often be modified by experience;[4] this is more likely the earlier the event in question takes place in the course of individual development. In infants a few months old, for instance, it is very difficult to distinguish what is innate from what is learned; and the difficulty is even greater where the vicissitudes of intrauterine development are concerned. In this case, experience is used not only via the nervous structure but also via other channels of information and transformation – for example, humoral channels.

1.4 Experiential structure and biology

A typical example of an automatic phenomenon determined solely on the biological basis of genetically transmitted structures is the patellar reflex. A light tap just below the bent knee of a person sitting with legs dangling gives rise to a rapid contraction of the thigh, causing the leg to jerk forward. This event is governed by a precise neuronal circuit. The tap impinges on the tendon linking the patella to the tibia and slightly increases its tension. Inside the tendon are receptors, made up of small cells differentiated in such a way as to be sensitive to mechanical energy: the stretching of the tendon triggers a biochemical process in these osteotendinous receptors that is propagated to the nerve fibre connected to the receptor. This fibre thus transmits a nerve impulse, which passes along the peripheral nerves of the lower limb to the posterior horns of the spinal cord; from here, the nerve impulse goes via a synapse to a motor neuron of the anterior horns, which transmits it to the efferent fibre, again via peripheral pathways, and hence to the quadriceps of the thigh, in whose neuromuscular plates the biochemical reaction of the nervous tissue triggers the biochemical reaction that gives rise to the muscle contraction.

The salient features of this phenomenon, apart from its automatic character, are the fact that it is constant in all individuals and independent of individual experience. Although it is possible, with concentration, to avoid

4 Under experimental conditions, the histological structure of the cortex is found to vary in accordance with experience.

the leg movement, this merely involves the superimposition of another – inhibitory – mechanism, which is characteristically mental and cortically determined. Moreover, this is possible only in man. Likewise, if the knee is not tapped, but instead pricked, burned or hurt in some other way, a corresponding mechanism of an analogous kind causes the affected part to be retracted. Much of our motricity is based on these reflexes, but that is not to say that all phenomena that have become automatic for us are reflex activities. For example, to ride a bicycle, a series of muscle regulations must be learned and become automatic; otherwise we should not know how to move the handlebars, balance the weight of the body, and so on. Although automated, riding a bicycle is a skill acquired through experience; it can be done only if it has been learned.

The above analysis of motricity applies, mutatis mutandis, to the senses; we ultimately perceive objects by psychic processing of the data from our sensory receptors. The field of perceptual phenomena exhibits many of the characteristics of learning. According to the reported experience of persons born blind who subsequently regain their sight, these subjects do not at first perceive objects as we perceive them. They fail to recognize not only the most elementary shapes but also the relationship between the size of objects and depth; they not only cannot recognize how a person is made, but also do not know whether something is big and distant or small and close by, because they have never had the experience underlying the acquisition of programs for processing data from the visual sensory channel.

Again, following sensory deprivation experiments (for example, being kept in complete darkness and absolute silence for several hours), subjects leaving the laboratory report substantial distortion of their perception of the external world. These phenomena suggest that the sensory mechanisms need either a background of experience or continuous use (stimulation) in order to generate the relatively constant psychic phenomenon of perceiving objects as they are and of distinguishing them from each other (Schultz 1965; Reed 1979).

Such experience might appear not to constitute 'learning' because it has virtually the same result for all subjects; it is therefore appropriate to distinguish between something that is acquired in the same way for all individuals of one and the same species, because conditioned by the morphophysiology of the receptors and brain of that species, on the one hand, and, on the other hand, a type of experience that determines variability between individuals. The latter is a different kind of learning, which belongs to the level of 'meaning'. The first type of acquisition, which is usually said to be necessary in order for the biological structures to become functional structures, is, however, also learned, and therefore not an innate factor. What is in fact learned in the same way by all individuals of a given species has too often been described as innate and hence biological. For example, a certain type of acquisition of visual functions is determined by the fact that our eyes are frontal and not lateral as in other animals, or because we have certain retinal pigments and not others; the fact that we are mammals entails an early tactile channel of relational learning; or, again, the fact of being male or female involves different primitive (mainly urethral) bodily experiences, which will underlie our subsequent sexuality (Imbasciati 1983a, 1987, 2000). It is quite difficult to define the boundary between the two types of experience – innate, or instinctive in the strict sense, on the one hand, and learned, on the other hand – and the difficulties increase the more primitive the stage of development under consideration. For example, in an individual in the first months of life, it is extremely difficult to distinguish between experiences that allow the normal acquisition of functions by the biological apparatus and those that determine their individual functioning.

In defining the mental, then, a distinction must be made between the genetic mechanism, which is rigidly determined by the chromosomal genes for each species, and the experiences that somehow enable it to function, either in the same way for all individuals of a given species or differently from one individual to another. Only genetic phenomena are necessarily innate; other functions are acquired and therefore involve 'learning', with the consequence of 'memory'. In the latter domain, the term 'innate' is often wrongly used to denote everything already present at birth; but this disregards the acquisitions of intrauterine life, as well as their importance in determining the biological structures themselves – in particular, the senses – in addition to the functional structures (the fetal psyche) that will underlie individual behaviour after birth and in turn condition the individual's subsequent specific ways of using all later experiences. Furthermore, the word 'innate' is often erroneously used in common parlance to denote everything that does not seem to have been acquired by *manifest* learning mechanisms – even though the connotations of the term 'learning' have been extended[5] to embrace any modification of behaviour resulting from a prior experience.

If the word 'learning' is used in this extended sense, the emotional structures that may be exhibited in the first weeks of life, for example, may be regarded as acquired; however, certain particular tendencies in sensory function that may be due to earlier (neonatal and prenatal) experiences can also be seen as acquisitions. From the fifth month on, the fetus has a neurosensory apparatus capable of receiving and processing many different kinds of information (e.g. auditory information); this processing activity leads to acquisitions, which may be peculiar to the specific individual concerned. Again, there is no need to assume a priori that acquisition takes place through neurological channels: consideration of embryonic development clearly indicates that characteristics may also be acquired otherwise than through the nervous

5 Through the notion of 'humoral memory'.

apparatus. This implies that the concept of a memory trace should also be extended so as to cover any situation in which an experience is recorded in the organism and the recording determines subsequent modifications of development.

The next problem is to distinguish what can legitimately be defined as 'mental' among the plethora of experiences 'acquired' by the organism. However, this problem cannot be solved without recourse to often questionable conventional antitheses – such as mental versus biological. For these reasons, the debate on the importance of innate versus acquired factors, which is already difficult in biology, sometimes becomes impossible in psychology. Whereas it is possible to determine that certain biological characteristics are innate, in the proper sense of the term, the same cannot readily be established for psychological characteristics; this does not mean that innate psychological characteristics do not exist, but only that it is not easy to verify their existence. Psychological science must therefore, as far as possible, establish which characteristics are acquired and how they come to be acquired. This is the background to our study of protomental processes and structures.

1.5 Innate or acquired? If acquired, when?

The study of the complex of functions that may be termed 'mind' was for a long time bound up with, if not dependent on, research on the nervous system – in particular, the central nervous system. Whereas it has long been acknowledged that the biological maturation of the neural substrate is not in itself sufficient, in the absence of experience, to produce functional capacities, the thorny problem remains of how this experience gives rise to functional structures, so that biological maturation itself ultimately proves to be conditioned by the acquired functional structures. Yet the very concept of 'biological maturation' is found to be ambiguous where the higher nervous structures are concerned. The so-called exercise of functions is known to be necessary for the establishment of functions. However, it is not easy to distinguish mere exercise – involving receptor stimulation, as required for the purely biological maturation of the neural tissues so that they become capable of performing their intrinsic functions – from actual learning, which is an acquired functional structure superimposed relatively independently on tissue maturation and on the functional maturation due solely to stimulation. Moreover, learning can modify the very morphology of the neural substrate, as experimental studies have shown (Oliverio 1986).[6] In a word, it is impossible

6 The cortices of rats subjected to intensive learning become thicker than those of rats raised in other conditions. Proliferations of fibres, synapses and glial cells are observed.

clearly to separate biological from functional maturation and functional maturation from learning. Learning involves the acquisition of a structure, which is admittedly functional – but also 'biological'.

A 'learning' operation comprises, first, the reception of stimuli from a sensory apparatus of extremely variable morphophysiology, each sense being specific to a particular kind of stimulus (for example, the retina to variations in electromagnetic waves, the cochlea to variations in pressure waves, the olfactory receptors to biochemical molecules, the thermal receptors to thermal variations, and so on), and, second, the 'reading' and processing of these stimuli and the preservation of the result of processing as memory. Hence learning is not the passive imprinting of data from external reality presented by receptors, but mental activity itself. This presupposes the existence of a neural structure capable of performing this activity; that is, one that has already acquired the functional capacities – at least that of elementary memorization – essential to any kind of 'reading'.

The following problem now arises: a functional structure capable of reading and processing inputs is necessary for learning, but prior learning seems to be essential to the acquisition of such a structure. What, then, is the starting point: experiential activities, on the one hand, or functional capacities innate (*genuinely* innate) in neural biology, on the other hand? The next problem is that not only external stimuli have to be received, read, processed and memorized, but also all the stimuli originating within the body itself, as well as the products of incipient mental activity itself. This activity, too, generates 'memorizable' products, which therefore enter into the functions and become capable of modifying the processes of reception, processing and memory; in a word, the first rudimentary thoughts produced by the incipient system may be said to enter into the construction and memorization of the first functions. Again, memory is never just recording that persists and accumulates, but a continuous process of transformation, applied to a huge variety of traces of 'information' of all kinds, each stemming from a variety of processing systems.

Clearly, then, a mental function of some kind must already be present if experience is to be used; in this sense, the psychoanalytic concept of 'learning from experience' (Bion 1962) is justified psychophysiologically too, in terms of reception, reading, processing and memory. One learns from experience on the basis of the types of functions that allow a kind of learning – and if the type of function differs from one individual to another, the learning in relation to one and the same experience will also be different. The memory 'trace' is not simply the nervous afference, from external or internal receptors, but something much more complex – something that can legitimately be described as mental.

What makes the problem of the processing and memorization of information even more complicated is that, whereas we have so far considered the mind in direct correlation with the neural substrate, information is also conveyed through humoral media, and there is no reason not to postulate corresponding forms of memory processing. The problem becomes even more acute if the beginnings of mental activity – in the fetus – are approached in terms of learning.

For example, a fetus at 6 months may be said to be capable of receiving a pressure stimulus through neural pathways, but it is impossible to decide whether a second-month ectoderm can in any way receive and retain any kind of stimulus and, if so, through what media. It is less and less possible to observe unequivocally mental events the closer the organism is to the beginning of life.

To avoid the impasse of as yet unverifiable hypotheses, and in line with our emerging definition of the mental, a conception of mind and an explanation of its origins that dispense with a direct reference to biology will be presented below. The mind will be considered as an information-processing system, irrespective of the form in which the information is recorded and encoded; and in this conception, 'mind' may be said to exist to the extent that the information processing achieved by the system at a given time can be described as an 'operation'. A computer is not a computer unless it performs an operation; the onset of the 'mental' lies not in the information itself but in the level of its processing – that is, in the presence of programs for processing the input received, as well as the consequent 'responses'. The term 'response' is used here not in the behaviourist sense, but to denote a kind of program, which may perfectly well be complex, that 'modulates' the individual, the modulation being evinced in or inferable from a behaviour pattern or from an 'observation' based on appropriate methods. In this last case, the 'response' may simply remain 'internal'; it then belongs to the very beginnings of thought, to which the psychoanalytic approach, and in particular certain contributions of the Bion school, will be relevant.

Obviously, therefore, any reference to the subject's consciousness will be useless. Again, in investigating the 'beginnings', we must consider the neonate, if not, indeed, the fetus. At most, it will be necessary to distinguish the forms of unconscious processing that can legitimately be described as mental operations, on the one hand, from the level of information gathering that does not qualify as processing – that is, as mental – on the other hand. In other words, are there forms of processing inconsistent with the implicit sense of the term 'operation'?

Since our intention is to 'explain' what psychoanalysis has described, we shall seek to translate the psychoanalytic viewpoint into a cognitive one, considering the individual's acquisitions from early relations with caregivers in terms of learning – that is, of the construction of memory traces of functions, or 'programs'.

Chapter 2

An explanatory theory for psychoanalysis

2.1 Description, understanding and explanation

Three levels of knowledge – descriptive, interpretative and explanatory – are distinguished in every science (Imbasciati 1994). The first two concern different degrees of *understanding* of the observed phenomena (*how* and *how well*), while the third relates to their *explanation* (*why*).

The first level calls for a consistent system for *observation of the facts* by means of appropriate instruments, whether these are technological instruments or logical ones that belong to the structure of the observer. *Observation* must be followed by appropriate *description*, which requires suitable concepts for describing the observed facts and conveying the observation unequivocally to other scientists. The concepts must therefore be defined as unambiguously as possible. The interpretative level needs *models* to connect the observed and described facts, even if the connections are only hypothetical; these models are not identical with the theory, but at most form part of it or facilitate its further elaboration. A theory, in the proper sense of the term, must cover the entire range of phenomena observed, described and partly interpreted by the models. Hence theory belongs in the continuum that extends from description and interpretation to explanation, and must somehow facilitate *understanding* of the observed field. Overall understanding makes for advances in the observational instruments and descriptive concepts, as well as in the interpretative models themselves; in other words, it possesses heuristic value for the progress of any science. The existence of a theory steers science toward explanation, even though many scholars consider that the underlying answer to the question *why* can never be forthcoming. Theories, then, are neither true nor false, but are more or less useful at a given point in the development of a science.

In this context, theory must not be confused with discovery (Imbasciati 1994: Chapter 5). The confusion of discoveries with theories is a fundamental epistemological error, of which many authors consider Freud to have been guilty. A particular error has indeed been handed down to us in Freud's oeuvre – namely, the failure to distinguish theory from discovery and both of these from method. Another epistemological confusion is that between the descriptive level of knowledge and explanation of the facts described. In the words of Holt (1972: 4f.):

> Certain methodological errors become habitual with Freud, and they often get in the way of our effort to grasp a precise meaning. He often reified concepts, even personified them, for he was fond of slipping into metaphorical, almost poetical usage. He did not always make clear distinctions between what he observed and what he assumed; in fact, conceptual innovations were often treated like empirical discoveries, and facts were confusingly mingled with theory. [. . .] He was so eager to discover basic laws that he often stated his observations with sweeping generality. He was also fond of extending concepts to the limit of their possible applicability, as if stretching the realm of phenomena spanned by a concept was a way to make it more abstract and useful. These propensities exposed him constantly to the dangers of oversimplification, which he tried to overcome by following one flat statement with another that qualified it by partial contradiction.[1]

Some of Freud's concepts are purely descriptive – such as conflict or resistance – whereas others have theoretical implications; for instance, 'drive' presupposes a 'dynamic' theory, or, in other words, a set of forces governing the individual. In Freud, this concept is incorporated in a theory that not only describes but also purports to explain, perhaps along organic lines, as with the supposed biochemical origin of the drives (Freud 1895d [1893–95]: 200ff.; 1905d: 167, 215 and 217; 1905e [1901]: 113; 1906a [1905]: 277; 1914c: 78; 1915c: 125; 1916–17a [1915–17]: 321). The difference between description and explanation is that the former reflects the occurrence of a certain observed sequence of facts (*post hoc*), whereas the latter postulates a causality (*propter hoc*).

On the distinction between description and explanation, and the ambiguity present in Freud, see B.B. Rubinstein (1967). The failure to distinguish between description, understanding and explanation is compounded in Freud by the confusion of concepts that denote a fact consistently observed and described (that is, 'discovered') with others that postulate causal links – that is, models and theories. The latter may help us to understand the observed facts, imposing unity and logic on them; however, if they are confused with the former, hypotheses and inferences risk being mistaken for established facts, so that the relevant statements become 'truth' and the theory assumes the status of doctrine.

1 It is perhaps this defect that led Popper to describe psychoanalysis as a non-science, on the grounds that it is not falsifiable.

A clearer distinction between discovery and theory is therefore necessary. For instance, the deceptiveness of subjective consciousness is a psychoanalytic discovery, and indeed a fundamental one, that has remained valid from the time of Freud to this day. The same applies to the phenomenon of resistance, as conceptualized in the descriptive models of the various defences. Free association, too, is a discovery, which underlies the method of psychoanalytic science; so are the contiguity and reciprocity of mental phenomena and bodily events, as well as, again, the importance of interpersonal relations and the highly specific patient–analyst relationship. The setting is also a group of discoveries, which are in this case methodological, while others are the relational phenomena called transference and countertransference, which are sometimes referred to by more sophisticated terms. Yet another discovery is the significance of actions, whether intentional or otherwise (from parapraxes via day-to-day behaviour to enactment).

However, do the so-called drives constitute a discovery? Or are they only a hypothesis, relating to a motivational model for the understanding of behaviour patterns? Or, as they appear in Freud, do they constitute a mechanistic explanatory model (Holt 1972)? The metapsychological concepts of ego, id and superego are likewise not discoveries but theoretical concepts. The concept of the internal object, too, is a synthetic way of describing and understanding a set of facts observed in clinical practice – but is it really a discovery? Or might it be only a means of facilitating the understanding of a certain kind of affective relationship? Again, is the 'Oedipus complex' a genuine discovery? Or is it just a way – and a mythical way at that – of describing triangular phenomena of jealousy and possessiveness with a view to accommodating them within a theory of sexuality?

In the development of psychoanalysis, confusion has reigned between discoveries and theories, as well as between concepts that stand for observable events, on the one hand, and, on the other hand, heuristic hypotheses for understanding and possibly explanation. This situation seems to have obscured the scientific foundation of psychoanalysis – namely, its method. Every science is based on its specific method, which characterizes the viewpoint from which it observes phenomena. In every science, theories change, discoveries persist and accumulate, and the method is increasingly refined. Freud is the founder of a method – that is, of a science – and not the discoverer of the Oedipus complex or of libido. The ambiguity perpetuated in Article 3(N) of the Constitution and Bylaws of the International Psychoanalytical Association[2] passes off theories as discoveries, thus presenting doctrine as truth.

2 'The term "psychoanalysis" refers to a theory of personality structure and function [. . .]. This body of knowledge is based on and derived from the fundamental psychological discoveries of Sigmund Freud.'

In the present-day psychological sciences, there are many theories of the development and functioning of the mind, which in fact constitutes the fundamental subject matter of the whole of psychology; almost all the successive theories of personality have included a basic scheme of this kind. In this respect, psychoanalytic theory undoubtedly possesses not only historical but also present-day validity. However, is it still meaningful to speak of psychoanalytic theory in the singular (Wallerstein 1988)? In any science, theory possesses only instrumental and provisional value, and so it is surely more appropriate to refer to psychoanalytic *theories* in the plural. This being the case, the theories in question must be identified (which is no easy task, given that the post-Freudian theories are cloaked in the penumbra of the Master's doctrine), and then those most relevant today, at the present stage of development of psychoanalysis, must be selected. In addition, and perhaps in particular, we should seek to integrate present-day psychoanalysis with current developments in other psychological sciences, which, while distinct from psychoanalysis, are concerned with the same object – namely, the mind.

This is appropriate principally if the scientific aim is explanation. The word 'aim' is emphasized here because the fundamental answer to the question *why* always lies somewhere in the future – to be pursued through the process of development by progressive approximation and reduction of the unknown that is characteristic of every science. After all, to a much greater extent than the descriptive and interpretative levels, the explanatory level requires the explanation to be consistent with the findings of other sciences concerned with the same phenomena from a different observational vantage point. Freud not only offered a descriptive and interpretative key to the *understanding* of psychic events (that is, one that permitted an understanding of subjectivity and its variation 'from the inside'), but also sought to explain them 'objectively' – that is, in a way consistent with the discoveries, or at least the hypotheses, of other sciences. This was achieved by his energy-and-drive theory (based on libido and drives), with its reference to instinct and a psychobiological energy. This was in line with the sciences of the time – in particular, with contemporary neurophysiology (the reflex-arc and electrophysiological-discharge models, etc.) and thermodynamics. These notions are not consistent with the findings of the present-day neurosciences, so that the Master's intended explanatory value can no longer be accorded to the Freudian theory. Again, Freud's energy-and-drive hypotheses underlie the part of traditional psychoanalytic theory that has been criticized for some decades. Each of the relevant critiques proposes a different model, but none of them, in my view, present a clear explanatory alternative to Freud's original theory. The energy-and-drive theory, after all, possesses heuristic value even today – that is to say, it can be useful for understanding the world of affects and hence for clinical practice – but it used also to have explanatory status, which Freud may have rated above its heuristic value, although it is now no longer tenable.

My own theory is based on a model consistent with the present state of the psychological and neuropsychological sciences, while firmly grounded in the mainstream of psychoanalysis, and offers an alternative explanation to that of the Freudian theory. It is therefore a psychoanalytic theory that can 'explain' development without resorting to Freud's energy-and-drive hypotheses. I hope that this will not collide with the prejudices of any over-deferential custodian of orthodoxy. Another aim of my theory is to build a bridge between psychoanalysis and the other psychological sciences – in particular, cognitive psychology. Hence my use of the term 'psychoanalytic cognitivism' (Imbasciati 1994). In my view, the psychoanalytic method possesses an exclusive specificity of its own, but a psychoanalytic theory cannot survive unless it is integrated with neighbouring sciences' theories of mind.

As the method characterizing the specificity of a science, I consider that the psychoanalytic method, which has advanced enormously since Freud's time,[3] is the one that has hitherto proved most suitable for (a) understanding the subjectivity (internal world) of the individual; (b) intervening in that subjectivity, allowing the subject to 'understand', reconstruct, or construct and improve himself; (c) understanding how a given adult individual can develop out of a given child; and (d) attempting to explain the development and functioning of the human mind. For these reasons, I believe that my choice – taking psychoanalysis as the basis for establishing a new explanatory theory – can be justified. Moreover, my work since that choice has enabled me to take stock of present-day psychoanalytic theories (Imbasciati 1978a, b, 1983a, c, 1989a); to consider the uncertainties inherent in them (1991); to review some of the main critiques (1994); to define clinical psychology (1993b, 1994), showing how it has impressed its stamp on the methodological principles of psychoanalysis; and finally to develop a psychoanalytic theory consistent with the experimental psychological sciences.

Again, the setting aside of the edifice of the energy-and-drive theory surely leaves psychoanalysis in need of an explanatory support – both as a foundation for clinical practice and to fill the gap, to which many authors (e.g. Klein, G. 1976) have drawn attention, between so-called clinical theory, whose validity is undisputed and which has advanced smoothly and continuously since Freud, and 'formal theory', which remains tied to Freud's explanatory hypotheses even though these have been rejected as such, as no longer compatible with the development of other sciences. A *theory for psychoanalysis* is therefore necessary – one that is not only psychoanalytic but also consistent with other sciences.

3 If a psychoanalyst were to work today in the same way as Freud – even the late Freud – did with his patients, he either would not be accepted as a member of a psychoanalytic society or would be struck off. Aircraft are not built today in the same way as they were in the days of the Wright brothers!

2.2 The explanatory aim in psychoanalysis

Freud's theory, considered as an explanatory theory of development, can be described in outline as follows: a force of natural origin – that is, one inherent in man's biology (libido) – impels human beings toward satisfaction; satisfaction constitutes the natural discharge of energy; in obtaining satisfaction, and discharge, the individual encounters, or comes into conflict with, reality and adapts to it (cf. the concept of adaptation [Hartmann 1939]); the energy is ramified and distributed into a number of different 'drives' that 'cathect' the various elements of reality, which thereby each assumes its specific psychic value; and a drive dynamic and economy, with cathexes and anticathexes, anxieties and defences, determine the development of every individual's psychic structure. The original scheme can be reduced to the following: *(biological) need + environmental frustration → defences → psychic structure*; or, more simply: *nature + environment = individual*, a relation that parallels *genotype + environment = phenotype*.

The 'reality' to which psychoanalysis refers is not merely an interaction with an inanimate environment, but the complex integration of interpersonal relationships, starting with that between the infant and the parent figures. Hence the concept of the 'object' and all the ensuing development of object-relations theories. These were devised and used to describe and understand *how* the individual psyche is structured; as for the question *why*, it was either set aside or referred back to the Freudian tradition – that is, to the hypothesis of an 'energy' modelled by certain 'realities', which deployed a dynamic and economy that led to the psychic structure of the individual.

Many authors have developed object-relations theories without reference to the energy-and-drive model, albeit without repudiating it; the entire British school belongs to this tradition. The developments of the Bion school, by emphasizing the aspect of learning *from* experience, imply that the structuring of the mind does not require endogenous forces (such as libido or drives), but takes place through learning processes, whose laws do not necessarily prove to conform to the Freudian paradigm, or to be susceptible to accommodation within the framework of the energy-and-drive theory. Even the concept of aggression, which in the work of Melanie Klein appears to be linked (in my opinion, only formally) to the concept of instinct (the death instinct), is subsequently detached from the instinctual paradigm (see Money-Kyrle 1955, 1968), and is replaced in the Bion school by the notion of destructiveness – and this, stripped of explanatory connotations (the question *why*), is used to describe a relational modality (*how*), which is based on the concept of fantasy rather than that of the drive.

The developments of object-relations theories have given rise to substantial divergences within psychoanalysis, with the result that many authors have attempted to combine the two models (drive and object). Typical examples can be found in the work of Kohut (1971, 1977), Gedo and Goldberg (1973) and Modell (1975a, b). Other authors, finally, have explicitly denounced the energy-and-drive hypotheses on which much of the Freudian theoretical edifice rests, rejecting metapsychology in particular; cf. George Klein (1976) and the review by Eagle (1984).

Attempts to reconcile the two models by superimposing object-relations theories on the Freudian theoretical structure are usually based on the assumption that the clinical facts determine which of the two models is the more useful. According to this view, a model and a theory are 'instruments' for understanding the clinical situation and for consequent modulation of the treatment, so that they are as such neither true nor false; it is the operational criterion that determines their adoption. In terms of theoretical relativism, the problem would thus appear to be solved. Yet the two models and the respective theories of mental development are incompatible: the former (Freud's) retains an explanatory value (*why*), while the latter seemingly concentrates on *how*. The ambition to explain, in authors who use the second type of theory, remains in abeyance, referred as it is back to the earliest object relations. These relations, not only neonatal but also fetal, are considered to determine the first structures,[4] which in turn are held to condition the subsequent processes of 'learning'. Hence the relevant authors' interest in research combining psychoanalysis, observation of the newborn and the fetus, and experimental psychology. A typical example is Stern (1985).

However, referring the situation back to the earliest object relations, or to neonatal and fetal experiences, does not explain the ultimate question (*why*) concerning the beginnings of a functional, psychic structure that is able to develop within itself all its subsequent increasingly complex functions. We *understand how* the first object relations (including those of the fetus) condition the acquisition of the first functional capacities, but no *explanation is given of why* an experience that is initially the mere reception of stimuli is followed by the capacity to process these stimuli. A biologistic explanation seems naive, even for the fetal psyche.

In my opinion, Bion's thought seeks to identify a *why* of this kind (concerning the transition from the biological to the psychic level and hence the beginnings of learning processes), given his specific attention to the forms of transition between a non-mentalizable sensory experience (sensoriality or sensuousness – cf. beta elements) to an experience that can be used to construct the 'mind' – that is, to the factors that generate the capacity for 'learning from experience' (the transition from beta to alpha).

It seems to me that Bion's interest in the transition from sensoriality to thought parallels the motivation for the experimental psychologists' research

4 The term 'structure' is not used here in a biological sense, but denotes an acquired functional structure.

on the origins of perception, in its distinction between sensory elements and the organization entailed by the perceptual act: in experimental psychology, sensory reception (sensation) was for a long time confused with and superimposed on the capacity for 'reading' and organizing sense data. Only this last event constitutes 'perception', which is a process different from mere reception, for it is an active process that, as such, presupposes acquired functional capacities. Nowadays, the perceptual process is regarded as one of the first 'mental' events. But what is the nature of the transition from the organization due to the morphology of the sensory receptors to the perceptual organization? The former may be attributed to the constitution and maturation of the receptors and neural pathways, but the latter entails a functionality (reading of input on the basis of a prior memory) that presupposes previous learning: can this be explained merely by biological maturation? Or does it presuppose other functions, also acquired by experience? In other words, does it presuppose that a mind already exists? This echoes an age-old philosophical dilemma (as expressed by Leibniz and Berkeley: *nihil est in intellectu quod non fuerit prius in sensu*, to which the answer was *nisi intellectus ipse*). If this *intellectus* is not presumed to be instilled or inherent in nature, its origins call for explanation. Nowadays we study fetal psychic life, but the problem of explanation remains – displaced on to the question of how (and possibly why) the first *mental* capacities arise *out of* fetal *experience* (Manfredi and Imbasciati 1997).

What, then, initiates the process whereby the neural substrate begins to acquire functions, thus enabling the progressive construction of functions that will ultimately constitute the mind to begin – and in such a way that the mind in question becomes capable of learning?

The cognitive sciences are developing so-called constructivist theories of the mind, which, it seems, can overcome this impasse. They postulate that the mind constructs itself through the progressive structuring of functional acquisitions on the basis of experience ('social' experience, as emphasized by some cognitivist authors, or primary 'object relations', according to the psychoanalysts), in a cause-and-effect circuit involving experience and neural structures. Kelly (1955) may be regarded as a precursor of this position; among more recent authors, we may mention Maturana and Varela (1980), Watzlawick (1984), Gilli and Marchetti (1992), and Camaioni (1993). According to these models, the affective-emotional structures are basic cognitive schemata (cf. Plutchik 1980) acquired at the very beginnings of life. These conceptions can be linked to the ethological models of imprinting, to the ethological-psychoanalytic theory of Bowlby (1969–80, 1979) and to the developments of Lichtenberg (1989). Present-day psychoanalysis, in general, traces the beginnings of the structuring of mental capacities back to the earliest object relations, both neonatal and fetal (Mancia 1980; Piontelli 1987; Negri 1993; Nathanielsz 1994). However, this does not explain why the acquisition of these capacities begins; many *how* questions are answered, but the *why* remains uncertain.

Freud's energy-and-drive theory had an explanatory aim (it was intended to answer the question *why*), and assumed the relevant energy to be biological in nature. As stated above, he postulated the existence of a biochemical substrate for the drives (see the Freud references given earlier in this chapter). Although this hypothesis was presented with due caution, it was taken up enthusiastically; indeed, Freud himself hoped that biochemistry might one day supersede psychoanalytic therapy. The idea of a biological, or somehow 'natural', underlying explanation remained for a long time implicit in the psychoanalytic mind, and after Freud's futurological speculations about a biological substrate were abandoned, the concept of the drive and the entire energy-and-drive theory continued to be used for many years. Explicitly, these concepts were employed in a metaphoric sense, owing to their undoubted heuristic value for clinical practice, but, in my view, the persistent emphasis on them may be implicitly attributed to their explanatory fascination – their connotation of *why*. The answer is held to lie in the 'natural' field, and its emergence has been postponed indefinitely, thus, in my opinion, deterring psychoanalysts from seeking an alternative to it. As a result, despite all the criticisms levelled for decades at the energy-and-drive theory, so that object theories have come to the fore instead, the latter have not been developed to the point of assuming explanatory value, apart from their facilitation of understanding and their clinical utility.

Yet it seems to me that object-relations theories, together with present-day developments in other psychological and neuropsychological sciences, hold out the prospect of also identifying the *why* – that is, the underlying reason for the commencement of psychic development at the beginning of life. This is the background to my 'theory of the protomental', whose foundations lie in some of the ideas of Bion.

2.3 Object relations: how do the mental functions come into being?

Inherent in the work of Bion (1962, 1963, 1965, 1967, 1970, 1974) and F. Bion (1984) are three intuitions on mental functioning and the origins of development that, in my view, possess the status of discoveries, transferred from clinical practice to theory. Stemming from the method of the Kleinian school and the theoretical models of object relations (cf. Imbasciati 1994, on 'method', 'discoveries' and 'theories'), they can be summarized as follows:

1. The affects, which are deep, unconscious entities rooted in the infantile world, are the basis of thought. They are themselves thought. Hence the cognitive processes stem from the affects.
2. The affects are made up of fantasies (in the Kleinian sense) that involve internal objects. Both the fantasies and the internal objects arise out of the child's earliest relationships through complex processes of internalization,

which have led to the constitution of an 'internal world'; on this basis, the individual develops his links with the persons around him, his relationship with the real world, his learning from experience, his individual psychic structure, and hence his capacity for knowledge.

3 The internal objects (the relations between which are fantasies and hence affects) are constitutive elements of thought – that is, of the way in which the individual comes to know the world. The external world is known through the internal world and through relationships.

The first two of these conceptual nuclei confirm once and for all that the affects cannot be conceived as bound up with an entity, such as libido or a drive, deemed to be a 'substance' different from what constitutes cognition. That disposes of the dichotomy of cognition and affect, which Freud had sought to establish and explain by the complex relationship between libidinal cathexes and known real objects – taking it for granted that these objects were perceived and represented automatically on the basis of brain structure, unless modulated by cathexes (cf. Imbasciati 1991, on representation and the constancy principle). This brings us to the third intuition – namely, that cognition (of the external world, as well as, on a more complex level, of the subject himself) is mediated by the internal objects. Here Bion follows Money-Kyrle (without, however, mentioning him), who had stated more explicitly that the internal objects are the means whereby a child (rather differently from an adult) represents the world to himself and thereby constructs the representations and, later, the concepts with which he apperceives and knows reality (cf. Money-Kyrle's 'concept building' [1968]).

Since our ability to know the world depends on the possibility of representing it to ourselves more or less appropriately, the problem arises of the 'representational value of the internal objects' and of the relationship between internal objects and representations proper – that is, the representations that can also gradually assume the clear and precise character which permits conscious thought. A continuum may reasonably be supposed to exist between the representational function of internal objects and representations proper – both in the diachrony of infantile development and in adult unconscious thought processes. Bion's grid implies the existence of this continuum.

Infant observation and child analysis seem to confirm this continuity. It extends from protorepresentations that have nothing to do with real objects (an infant represents the world to himself in this way, through internal objects that bear no resemblance to any reality), via others in which an approximate correspondence begins to emerge, to representations proper. There is in effect a chain of signifiers that reflect reality with progressively better accuracy. Notwithstanding the existence – or rather the construction – of signifiers that afford increasingly appropriate representations of reality, the earliest signifiers persist and retain their intensity: the primary internal objects remain (and operate) even after actual representations have come into being. The existence of a continuity between internal objects and representations does not invalidate the psychoanalytic concept of conflict, which can be described in representational rather than energetic terms as a discrepancy and contradiction between signifieds along the chain of significant representations (Imbasciati and Calorio 1981; Imbasciati 1983b). This corresponds to Bion's inner 'lie'.

Bion also confronts head-on the problem of the transition from the sensory to the mental level: the question of course arises as to what it is that initiates the process whereby the particular 'something' that can be processed and acquired – and hence memorized – comes into being out of the neurological afferences due to the morphophysiology of the receptors. This raises the problem of conceptualizing these acquisitions in terms of memory traces; at the same time, the most important memory traces are manifestly those of acquired functions – traces of programs acquired step by step and constituting progressively greater capacities. These could be described as capacities for thought, forming the basis of all subsequent 'learning from experience'. Bion seems to have solved the problem of the onset of the earliest learning and the constitution of the first mental capacities by his concepts of alpha elements, beta elements, beta screens, the alpha function, etc. In my opinion, behind Bion's abstract and mathematical model lies the same problem as that addressed by the experimental psychologists in their study of the process of perceptual organization, and in particular the concept of representation as the mnemic foundation of the 'reading' that gives rise to perception (see Section 3.4) and the subsequent development of learning. Perception proves to be the psychic process on the basis of which the other cognitive processes can take place, but there still remains the problem of how the functional capacities that give rise to the first perception are acquired. In other words, Bion's question addresses the problem of the transition from the morphophysiology of the neural tissues, developed by the genetic program, to the functionality that may be called 'mind', as well as to the constitution of the relevant traces in the neuronal tissues.

Bion's approach raises three questions pertinent to the search for the *why* of psychic development, which my theory seeks to place in a framework that takes account of the current status of the experimental psychological sciences. The first concerns the genesis of the internal objects. It is true, if simplistic, to say that they stem from relationships; but we must also describe, and then explain, how a sensory afference, which, after all, is the physical medium of all communication between living beings and hence also of any interpersonal relationship, becomes an internal object – how it is established as a trace and what is the nature of this trace. This leads to the second question, concerning the possibility of conceptualizing primary learning as the transition from purely sensory information (input) to mental organizations possessing representational status (internal objects first, and representations proper later); in other words, how are we to conceptualize, in more detailed terms than Bion's abstract models allow, the more general transition from the senses to mentalization? The third question, ancillary to the first two, relates to the possibility of analysing the perceptual process in terms of experimental psychology and psychophysiology, the analysis at the same time being consistent with and usable in the context of psychoanalysis; in other words, what is the relationship between the genesis of perception and that of the internal objects (see Sections 3.2 and 3.3)?

In addition to the above three questions, which arise in the further development of Bion's theory, a fourth concerns the possibility of constructing a psychoanalytic theory that includes an 'explanatory' aspect and is thus compatible with the present-day findings of other sciences. In Freud's day, the neurosciences used terms compatible with the concept of energy and its transformation – cf. Freud's 'libido' and 'drives'. Today's experimental sciences posit that the mind entails the capacity to process its inputs in such a way as to give rise to an activity capable of 'reading' experiences – which, we may add, include internal as well as external experiences. My theory seeks to conceptualize the experiences stemming from object relations in these terms.

There is in fact also a fifth and final question, which can be developed only partially and hypothetically here since the relevant literature data are contradictory. With regard to primordial states of the mind, many authors have invoked pre-object stages (Bleger 1967; Bollas 1987, 1992; Ogden 1990) or pre-representational stages (Greenspan 1997), but is it legitimate to adduce experiences and the relevant memory traces derived from relationships predating object relations? Or predating representations? Is the concept of an object relationship being detached from that of a relationship in general? Infant research shows that memory traces, of both content and functionality, already form in the fetus and the neonate. It follows that, regardless of the psychoanalytic method's dating of the 'object' (which may be an internal or part-object) and of an 'object relationship', there is an experience which cannot be anything but relational and which leaves traces. The concept of the internal object must thus be reformulated in different and less ambiguous terms, and so must, in particular, the concept of a representation. The term 'representation' need by no means be used solely in the sense of a representation that matches reality or real objects. Again, whether or not the relationship between a child and his world can be described as an object relationship, the incipient mind of a neonate (and, still earlier, of a fetus) enters into a relationship with something (central to which is the caregiver) whereby it processes an experience, with the relevant memory traces. It is on the basis of these memory traces that the child gets his bearings in the world and relates to the caregiver; he therefore has *some kind* of a representation of the caregiver. This study therefore seeks to develop a mnemo-representational theory that does not use the term 'representation' in the restrictive sense in which it is employed by some psychoanalytic authors.

If the energy-and-drive hypothesis can no longer be invoked to *explain* the origin of psychic life, an alternative theory is necessary. That psychoanalysis is tending in this direction is, in my view, suggested by a number of diverse studies by various authors. Many see responses, which may be complex, to specific configurations of stimuli as the basis of a 'mental' activity whereby successive learning operations will become possible. Ogden (1990), for example, uses the term 'instinct' (in my view, incorrectly) to denote an innate code for reading certain experiences: in his opinion, a functional organization is acquired with the appearance of certain experiences, but depends on the presence of a biological predisposition. This may be likened to Bion's 'pre-conception'. To explain the transition from a mental activity of this kind, which is, by definition, incapable of 'reading' reality, to one with the ability to learn, Ogden invokes the mechanisms of projective identification.

It seems to me that, while projective identification can help us to understand the *how* of the process of genesis of the mind, it tells us little about the *why*. For this purpose, we must consider the transition from sensory afferences (that is, of experience – including, moreover, internal experience) to the capacity to process them, *from* the experience itself. This has to do with the capacity to organize the multiplicity of individual inputs into operational units that will make mental 'operations' possible. My theory is directed toward these problems, starting as it does from relationality and presenting a model in which the source of experience can be conceived in terms homologous to known entities and phenomena such as afferences, inputs, processes, reading or the memory of reading.[5] I hope that my theory will eventually be integrated within and indeed superseded by the work of other authors.

My theoretical framework takes sensory psychophysiology as its starting point for the explanation of psychic development from its beginnings in terms of psychoanalytic concepts (Imbasciati and Calorio 1981). This implies attributing not only heuristic and clinical but also probative value to these concepts for the purposes of 'explanation'. The model is constructivist in nature and takes account of some elementary principles of cybernetics. The brain may be likened to a huge computer, with the mind as its set of stored, or 'learned', 'programs', or functions. The overall capacity of the computer varies with the variation of these functions: psychic structure is the totality of these functions, which differs from one individual to another. Initially, the

5 My model is empiricist rather than innatist in nature, emphasizing as it does the value of experience over that of biology. This is not to underestimate the importance of the neural substrate, for it seems to me that my position is consistent with recent studies demonstrating the complex feedback processes between morphology, physiology, learning and hence also between physiology and morphology themselves; even neural morphology is found to be modified by learning (Oliverio 1986).

most elementary functions are learned. This initial memorization of functions, which constitutes the primary nucleus of the mind, modulates all subsequent learning and all subsequent and progressively acquired memory, as well as the memorization of new functions and contents, all of which are inseparable from each other. In this way, the mind learns and at the same time constructs itself – that is, it constructs its own capacity for learning.[6]

All learning depends not only on the capability of the hardware (the brain of *Homo sapiens*), but also on the input or sets of inputs; in other words, the structuring of what is memorized must have its origins in sensory information. This is readily conceivable in the case of an already developed mind (one that is, so to speak, 'up and running'), but becomes problematic at the nebulous level of the genesis of the processing operations known as 'affects'. This is partly because of the need to come to grips with neonatal and fetal sensoriality, but mainly because we need to explain how the first memorization takes place, given that memorizing depends on perception and that, in order for there to be perception, psychic processes of recognition, reading and hence prior memory are required.

Neurosensory input itself, considered as an isolated entity (the individual stimulation of individual receptor cells), is not retained in memory; it neither enters into the biochemical memory laid down in RNA, nor is it preserved for any length of time as a bioelectric trace. The input must be 'read' in order to be memorized, and the reading depends on recognition of the 'sets' of which it is composed. For a 'set' to be 'readable', there must therefore be a corresponding memory *trace* constituting the equivalent reading unit; and all this must correspond to the neurophysiological events that underlie perceptual recognition, or rather the 'reading' of the elements that will give rise to perception.

A number of questions now arise. How are the first reading units formed? Which input 'sets' will make up the first memory? Are there sets that are memorized as such and can therefore constitute the first reading units? If not, if reading units are also needed for the first inputs, how can the first 'function' allowing the first memorization commence? There is a problem of 'assembly' – of the individual neurosensory inputs – which either allows or does not allow the traces of these inputs to be memorized. Why and how are these traces memorized? What are the characteristics of these first assemblies of traces of sets, which therefore constitute the first protorepresentations and hence the first perceptions? What are the modes of assembly, of both inputs and their traces, that allow the constitution of initial 'functions', which will in

6 When confronted with words like 'learning', 'reading' or 'memory', which are not usually applied to the affects, some analysts may feel removed from their psychoanalytic element. While that element is the essence of the understanding of subjectivity, other concepts, too, are needed for the purposes of explanation, and will not be harmful.

turn permit the first learning operations? Are there different, graded orders of protorepresentations? What are their respective modes of assembly? All these questions can be posed in neuropsychophysiological terms, identification of the corresponding traces being their explanatory support, but can they also be expressed in equivalent psychological terms – for instance: What does a neonate perceive? What does a fetus perceive? What representations (or rather protorepresentations) must a neonate (and perhaps also a fetus at an advanced stage of gestation) already necessarily have in order to be able to perceive? How have they been formed? What type of learning from experience has occurred and what type of learning does it permit? How are so-called affects formed?

Although neither a neonate nor a fetus possesses consciousness, their perceptual processes must be considered and the relevant protorepresentational engrams that allow perceptual reading must be sought. The adult unconscious can now be studied in the same way. Bion's grid helps us to see perception and other conscious processes as no different in nature from unconscious processes. What, then, is the value to psychoanalysis of all the above questions, whether formulated in explanatory terms of neurophysiological traces and their processing or in descriptive terms in accordance with the observations of experimental psychology? Which psychoanalytic concepts are superimposed on and explained by them? Take, for example, the internal objects, to whose protorepresentational value for the first perceptions (that is, knowledge) of the world I have drawn attention before (Imbasciati 1991), and do so again here (see Sections 3.2, 3.4 and 10.2): what do the internal objects correspond to in terms of orders of protorepresentations and hence of the assembly of traces of inputs from various senses? Do these internal objects underlie a certain type of perception rather than of hallucination? Moreover, when psychoanalysts maintain that the absent real object is perceived as if it were a present bad object, how does a neonate perceive the inputs whereby this 'bad object' presents itself to him – that is, is perceived by him? It is not merely an object of affects, but also a perceptual object (Imbasciati 1993b). If a neonate is frightened by an unusual noise, this means that the noise has been perceived (that is, read, by protorepresentational reading units) as a bad object. What protorepresentational reading units have been deployed? What traces from what senses have contributed to the construction of these units? Again, when a baby transforms the enteroceptive afferences of hunger into the perception of a bad external object, what kind of protorepresentations are involved? There is, then, a continuum between hallucination and perception, and representational units are required for each step along it.

Furthermore, psychoanalysis (and Bion, in particular) holds that the paranoid-schizoid mechanisms do not allow internal reality (bad objects) to be distinguished from external reality, so that the latter is perceived as threatening and different from actual reality. What kind of protorepresentations are involved in such mechanisms and the consequent perceptions (or hallucinations)? What assemblies of external and internal traces are 'mixed' in such a way as to give rise to confusion between perception and hallucination? In adult unconscious processes involving the Ps ↔ D oscillation described by Bion, what 'metabolism' of protorepresentational traces takes place in the Ps (paranoid-schizoid) phase? Similarly, what changes in assemblies of protorepresentational traces parallel the shift to the depressive phase? Is this last 'reassembly' the factor that allows the emergence of a reality-consistent perception and hence of the conscious dimension of perception itself?

In other words, we are concerned here not only with the outside – how external objects are perceived – but also, and in particular, with the inside. In this respect, our theory claims to be psychoanalytic: just as sensory inputs are processed by an acquired functional structure that gives rise to cognition of the external world, so we shall inquire into the origins of the plethora of internal products that psychoanalysis has come to label as affects, drives, conflicts, internal objects, defences, etc. (wrongly supposing them to be endogenous, as Freud did) and how they are processed, until they become distinguishable from what is read as coming from outside. It will also become possible to distinguish between information from the outside world and internal manifestations. This corresponds to the capacity to represent one's own internal states to oneself, as described by Greenspan (1997) – including the functional structures acquired in the course of individual psychic development and the history of the individual concerned, which characterize that individual's unconscious internal world. In that world too, there may be cognition: knowledge of self, which may be non-conscious, distorted and even mendacious, but which will nevertheless be a way of connecting with the information circulating in the internal 'mind system'. After all, these forms of connection will determine the quality of the most complex (interpersonal) relations with the external world and hence the way in which this world comes to be known. The type of interpersonal integration on which the entire progressive construction of the personality will depend is intrinsically determined by a way of knowing – of knowing oneself, others, the world, and the second and third of these through the first. The structure of the unconscious thus proves to be the foundation of all knowledge.

2.4 The mind system

By its characteristic emphasis on the unconscious internality of psychic processes, psychoanalytic research risks reinforcing a popular notion that psychoanalysis itself in fact set out to disprove – namely, that the human mind consists of a 'psyche', understood as a kind of mysterious substance that moves human beings and develops by itself.

Popular culture has indeed embraced the psychoanalytic concepts that suited it best. The emphasis on development of the mind by endogenous instinctual forces and on the unconscious has reinforced the notion of a psyche whose origins and processes, being removed from conscious control, are commonly, but wrongly, supposed also to be removed from the experience of external reality – from the experience which, in the popular imagination, has to do with the control of consciousness. This aspect has tended to confirm another prejudice – namely, that of a mental sphere of rational and conscious cognitive processes, separate from a deep, unconscious 'psychic' sphere (here the adjectives 'psychic' and 'mental' connote the difference). In other words, it has lent weight to the old prejudice inherent in our Western culture of an affectivity separate from cognition. This separation reflects the distinction, rooted in an older tradition, between a concupiscible and a rational soul, which can be contrasted with each other on the basis of the contraposition of two different substances, body and soul, as presumed substrates of the two aspects of psychic functioning.

While this error has not taken hold in the science of psychoanalysis itself, it is contained in the image of that science absorbed into general culture. However, it has to some extent been fuelled by certain aspects of Freudian psychoanalytic theory, and in particular the endogenist emphasis on libido theory. The work of Bion, with its stress on the concept of 'mind' and on learning from experience, is now helping to correct this error. Indeed, the present theory of the protomental takes some of Bion's assumptions as its starting point.

Underlying this theory is a general model of the mind that is substantially[7] cognitivistic in nature, in which the mental apparatus is seen as a computer-like data-processing system. The theory of the protomental seeks to explain the origin and functioning of the 'mind system' in semantic-representational terms. In contrast to the Freudian energy model, it accommodates what the object-relations theorists have conceptualized in terms of relationship, experience and learning from experience. From this point of view, it constitutes an alternative to the Freudian theory.

A number of authors, especially in recent decades, have pointed out that Freud's energy-and-drive theory is separate from, if not inconsistent with, his clinical theory. It has gradually been realized – for example, through the work of George Klein (1976) – that Freud's oeuvre contains two theories, a clinical theory and a metapsychological theory, and that only the former remains valid, the latter being an explanatory excrescence (Klein, G. ibid: 41ff.) devised by Freud by analogy with the contemporary natural sciences. Object-relations theories are basically clinical theories; they are reconcilable with Freud's clinical theory provided that they are not also based on the

7 Substantially, in so far as it identifies the cognitive value of any psychic event; however, it differs radically from what is described by so-called cognitivism, in its commonly understood sense and with its usual connotations.

metapsychological energy-and-drive theory, although many authors (e.g. Kohut), in desperation and confusion, seem for many years to have been compelled to retain the latter. Faced with the untenability of Freudian metapsychology, or rather of Freud's explanatory theory, many authors (Fabozzi and Ortu 1996) opt to see only the clinical theory as valid for psychoanalysis ('metapsychology is not psychology' – Gill 1976); in other words, they accept only a theory of description and understanding (from the inside, in the subjectivity of patient and analyst), while forgoing explanation. Other authors hold that an explanatory theory is essential; these include Peterfreund and Franceschini (1973), who, like myself, adopt an information-processing solution.

The theory of the protomental is intended to offer explanation on the general interpretative level, incorporating the clinical theories of psychoanalysis – in particular, object-relations theories – within a general theory whose aim is not only to provide a support[8] for psychoanalysis, but also to be not inconsistent with the other (experimental) psychological sciences, or with other sciences concerned with the mind from a biological standpoint (neurophysiology or neuropsychology) or in terms of logic and cybernetics.

The mind can be conceived as a system of functions that acts on an input so as to produce an output, which consists of all observable behaviour patterns and the processes, or operations, that underlie and may influence them. The following must be analysed in this system: (1) the quality of all possible inputs; (2) the input-reception system; (3) the processing (that is, encoding) of the received data; (4) the storage of the processed received data and its possible subsequent transformation; (5) the processing system, connected to both the reception and the storage systems; (6) the quality of the functions determinable in the processing and identification of the processes leading to the output; (7) the variety of methods that can be used to observe these processes – that is, the viewpoints of the various psychological sciences, including psychoanalysis, used to identify and determine the processes; and (8) feedback phenomena between the various subsystems described above.

In Figure 2.1, 'I' denotes the *input* made up of all the information arriving from the external world – that is, all the modulations of various types of physical energy that can act as *stimuli* (S) in the different sensory *reception* systems (R). The principal modifications of some of the forms of physical energy that can act as stimuli in human beings are as follows. First, there are modulations in the intensity of electromagnetic waves in the range 3500 to 7000 angstroms, which stimulate the population of retinal receptors, thus triggering the *afferences* (arrow A) whose processing generates the perception of

8 In any science, a theory is never either true or false, but is an instrument whose validity is determined by its potential usefulness for the development of the relevant science (cf. Imbasciati 1994: Section 3.5).

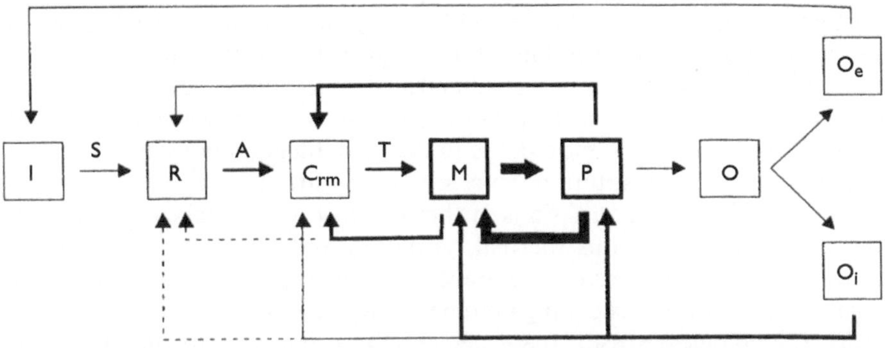

Figure 2.1 Logic diagram of the mind system. The input (I) generates stimuli (S) that can be received by receptors (R) and transformed by them into afferences (A). These are read and transformed into traces (T) by the reception memory code (C_{rm}). The traces can be stored (M = memory) and are continuously processed (P) and re-stored in a process of constant restructuring. The combination of M and P is the heart of the entire system. It issues programs (O = output) in two directions: (1) to the outside (O_e), in the form of mental 'products' for updating the subject's various behaviour patterns (motricity, expression, communication, etc.); and (2) to the inside (O_i). The latter group of products of mental processing remain within the thinking system itself (feedback) and modify its functions and storage. The diagram also shows other functional feedbacks, the thickness of the lines indicating their intensity.

light. Still in the field of electromagnetic energy, frequency modulations give rise, again in the retinal receptors, to different and corresponding biochemical reactions in the various rhodopsins, whose relative proportions determine the processing of afferences into colour perception. Again, the respective spatial distribution of the first and second type of modulations within the field scanned by the eye's lens system gives rise to the type of retinal afference that is processed into the perception of shape, as well as to the more complex processing of afferences and efferences that underlies stereoscopic perception.

Other inputs (I) are modulations of the physical energy of compression and decompression of matter – namely, sound waves. Modulations of amplitude, frequency and waveform stimulate the corresponding coding operations and perceptual processing of the intensity, pitch and timbre of sounds and noises.

Next there are the modifications of mechanical energy, which, if applied to the surface of the skin or otherwise transmitted to the skeletal and muscular apparatus, can stimulate the cutaneous tactile receptors, the dermal pressure receptors and the osteotendinous and muscular receptors. These stimuli give rise to the multidimensional afferences whose processing in turn generates so-called proprioception, which lies at the root of the complex perceptual system known in the aggregate as 'touch', the final processing of which neurologists describe as epicritic sensibility.

Then there are the variations in the position of the body and its various segments, which stimulate the vestibular receptors and the associated neuromotor integration, the encoding of which gives rise to the subject's perception within the fundamental coordinates of space, with the corresponding continuous variations, and to the coordination of all movements, from the simplest to the most complex, down to the final integrational steps; all this is termed the motor constitution, body scheme, bodily self, etc.

Consider finally the variations in thermal energy or biochemical energies that stimulate the various thermal, gustatory and olfactory receptors, the processing of which gives rise not only to the corresponding perceptions, but also to non-conscious encoding (and hence not to actual 'per-ceptions' in the strict, etymological sense), these perceptions being of vital importance in terms of the processing they can initiate within the system as a whole. In this connection, for this and all the other sensory processes mentioned above, we are of course concerned not only with the perceptual result, in the strict sense – that is, in the subject's consciousness – but with the processing, in whatever form, that we observe or infer within the system as a whole.

Finally, 'I' includes inputs from the in-tissue biochemical receptors (enteroceptive sensibility).

Thus, 'I' stands for the entire range of modulations of the various physical energies that determine sensory reception, which are thus specific to the receptors of every animal species. Some animals are stimulated by modulations outside the human stimulation range (such as ultrasound), while others may be sensitive to types of energy (for instance, magnetic fields) for which no receptors have been discovered in man.

To return to the diagram, system I is thus translated into S (stimulus), which is collected by the receptor system R, from which afferences (A) issue. These are encoded, or processed, by the subsystem C_{rm} (*reception memory code*). The function of this subsystem (which cannot be identified anatomically because it is divided between the peripheral, subcortical and cortical neural stations) is to transform the afferences into a set usable for subsequent processing operations, describable as mental or 'psychic'. For an example of this gradual process of organization, see the detailed description of the progressive aggregation of afferences into a perceptual image (Section 3.2). Perception is by no means an automatic process whereby afferences are imprinted on some kind of sound- or vision-reproducing apparatus, but an active process of reading of the afferent data by the mind system, involving the aggregation of the data into sets with the status of operational units for construction of the perceptual result. Perception proper is the end result of a long chain of non-conscious processes carried out on the relevant afference. The processes leading to perception, inferable experimentally, can also be observed by appropriate analytic instruments, especially in dreams. For this reason, the non-conscious processes whereby afferences are processed may be regarded as unconscious from the point of view of various sciences, including psychoanalysis. Analysis of perception is an analysis of the reading operation carried out by the mind system on the afference, or, better, on the afference as processed into 'traces'. A trace is not the imprinting of an afference as such, but the result of processing of the afferent input – a 'reading', which is stored. Otherwise the afference, if not read – that is, organized in progressive codes – is dispersed within the system, and no trace of it is preserved at the neural level. The organization of afferences observable in the process that culminates in the conscious perceptual image is only one example (albeit the most obvious) of a more general process of organization of afferences that takes place whenever something is received, even if what is received does not give rise to manifest perception or learning processes (cf. research on subception and incidental learning).

Moreover, as may reasonably be supposed, the same applies to many other processes that take place within the system. Using the functions progressively acquired, the system 'produces' something that adds to and increases its previous functionality. These 'products', which become functions, have their corresponding traces: not only the afferences from external inputs, but also everything produced internally at all times by the constantly active system undergoes continuous reprocessing, and the two levels of processing are mixed and integrated as traces are constantly reprocessed. All the mental processes taking place in the system, with their respective operations as inferable from the functionalities that can be reconstructed at the various levels of the system itself, underlie the end results of processing, which are the 'mental processes' described, according to the nomenclature of the relevant psychological discipline, by terms such as images, representations, fantasies, memories, thoughts, feelings, dispositions, motivations, affects, internal objects, and so on. In all mental processes, we must presuppose, and can discover, a constant restructuring of the 'traces' used progressively for each of them; these 'traces' are complex functional (and neurological) elements stored as *functions*[9] in the system. The possibility of performing a given function – that is, of the occurrence of a given psychic process – depends on the presence of the

9 A trace should not be seen as the simple, direct residue of an afference, for the entity in which a certain type of function, process or response is laid down is also a 'trace'. Every 'product' of processing that has arisen in the system is laid down as a trace. Again, even in the case of traces concerning perceptual representations, it would be naive to believe that such a representation is a copy of what is subjectively perceived. Kaufman (1974) described this prejudice, which is not only stubbornly upheld in the popular mind but also widespread in scientific circles, as the 'picture-in-the-head theory' – in other words, the idea that the head contains 'figures', visual or auditory for the most part, that correspond morphologically to what is perceived. This notion is a residue of Köhler's (1929) theory of cerebral isomorphism. Hence a representation, as well as every memory trace, corresponds in the brain to units of biochemical (molecular) code that bear no morphological resemblance to the subjective experience of a perception or representation.

trace of that function. These traces also include ones referable to what is received and then processed by the senses, as observed when the progression of perceptual operations is analysed. This is manifested in the recall or construction of images expressible in perceptual terms (such as painted images, or a piano performance), but it is also inherent in abstract thought, whether reproducible through language, through spatial geometrical symbolization, or through mathematical or indeed artistic symbols. In other words, in the consideration of any mental process, operational units and their corresponding traces of functions cannot be dispensed with. The primary formation of both the former and the latter is initially bound to be of sensory origin. The affective processes are also linked to sense-related operational units, as is clear from the links between perceptual representations and the possibility of experiencing, expressing or communicating affects by projective identification,[10] or between affects – in particular, feelings – and memories. The internal objects described by psychoanalysis, too, as units used by a child to represent the world to himself, have a sensory reference – which, of course, has nothing to do with a perception in the adult sense, but involves some kind of organization of sensory reception (see Section 4.2). Finally, psychoanalytic communication, which is utterly pervaded by affects, uses the senses to convey them. Affects cannot be communicated without a sensible medium of some kind.

Hence, the term 'trace' has a very different meaning from that customarily assigned to it by late-nineteenth-century physiology. As we shall see in Chapters 3 and 4, in our theory the concept of a 'trace', understood in this extended sense as the trace left by the processing of information whereby acquired functional capacities are recorded, performs an explanatory function similar to that of the drive in Freud's theory: quanta of circulating energies are replaced by memorization of the results of processing of circulating information.

Our subsystem C_{rm} not only concerns perception, or the reception of afferences of one kind or another in the here and now, but is also a processing and reprocessing station that also modifies the results of processing carried out by the subsequent subsystems, which, as Figure 2.1 also shows, are looped back to C_{rm}. The theory of the protomental specifically sets out to describe how, by progressive encoding and processing operations, sets of traces that can be retained and used for subsequent processing operations are organized from the raw afferent data. The problem concerns the transition, described by Bion, from sensory to 'mental' elements – in other words, 'from the neurological to the psychic' or, better, from an input in the raw state to

10 Projective and introjective identifications, as well as counter-identifications, are not conveyed through mysterious parapsychological channels but are of course always mediated by sets of sensory elements.

operational units that can be laid down in memory and undergo further processing.

The arrow pointing to the right from subsystem C_{rm} in Figure 2.1 indicates that the primary processing operation carried out by the subsystem has enabled some kind of retention in memory – namely, a *trace* (T) – which will pass to subsystem M (*memory* or *store*). The existence of a trace depends on the reading system that transforms afferences by means of the reception memory code. For simplicity, subsystem M is deemed to include all the various possible types of memorization, from the volatile ones known to be connected with the flow of bioelectric currents in neuronal networks to those stabilized in biochemical modifications of (neuronal and glial) RNA and in the corresponding (albeit not yet biunivocal) functional levels of *short-term* and *long-term* memory (STM and LTM respectively). On both of these levels, it is possible, in so far as the methodology of certain psychological sciences and of psychoanalysis in particular allows, to distinguish conscious memorizations, memorizations that can be made conscious and ones that cannot become conscious; the last may be defined, according to one's chosen psychoanalytic theory, as defences or as primitive primary-process levels.

Subsystem M, which, as its name implies, can store its own input, is closely connected not only to subsystem C_{rm} (encoding of afferences), but also to a *processing* subsystem, denoted in Figure 2.1 by P. In other words, memory is anything but passive storage, and the mind is anything but an apparatus for reproducing external reality. A sensory afference in itself cannot be used by the mind or retained; before use and retention are possible, it must be transformed and processed. Here we may seek what Bion subsumed in the notion of the transition from the sensual to the mental. Subsystem M is therefore seen as intrinsically linked to the set of transformational and processing functions (which are therefore also creative) that are constantly at work in the 'mind system' and which, so to speak, constitute the very heart of the mind. This subsystem (which, for the sake of simplicity, can be identified with the system as a whole) is closely connected to the store; indeed, the connection is so intimate that it is perhaps impossible to distinguish it from the store itself; at any rate, M is in continuous circular interaction with the store. This inseparability and continuity are indicated in the diagram by the thickness and shortness of the arrow linking M and P and the similar thickness of the feedback arrow P $\rightarrow$ M. For the same reasons of inseparability and continuity, the two boxes P and M are drawn with thicker outlines. P and M constitute a single assembly, which is the central core of the mind system. The *processor* P is made up of all the processes that constantly use the stored data to create other data from it, to be subsequently stored or used for the processing of a given output, as shown below. This assembly, here named as a single entity, can readily be described in more detail: all psychological studies, irrespective of their standpoint and specific discipline, seek to identify and describe the processes, actual or presumed, making up this 'processor'. For instance, the study of cognitive strategies in problem solving involves a particular description of a functional aspect of this processor, in this case from the point of view of cognitive psychology. Similarly, in establishing the laws of operational conditioning, the behaviourists describe a particular aspect of the processor's functions in their own terms. Likewise, psychoanalysis, in accordance with its specific methodology and theories, describes various levels of psychic processes, defence mechanisms, structures and apparatuses, introducing for the purpose appropriate concepts and models – for example (in Klein or Bion) fantasy, paranoid-schizoid or depressive functioning, splitting, projective identification, evacuation, the alpha function, the beta screen, and so on.

The aim of defining all these processes is to explain how the processor produces a certain *output*, represented in Figure 2.1 by the next box, O. This output is usually deemed to include the subject's behaviour, or, better, patterns of behaviour. Some psychological disciplines – in particular, the behaviourist schools – consider this output predominantly on the level of its external and manifest aspects, and correlate it with the input by way of their explanation of the processing mechanisms in terms of conditioning, generalization, extinction and so on, all within the model of a stimulus–response circuit. This reductionist conception has now been abandoned even by the behaviourists, and the output is now considered to include, for example, language, patterns of verbal behaviour and hence also the thought inferable from these. Taking a wider view, we could define output as anything produced by the processor even if not transmitted outside the system. Two types of output (O_e and O_i respectively in Figure 2.1) can therefore be distinguished: first, an *output externalized* in the subject's patterns of behaviour, and, second, an *output* that remains *inside* the system as a new creation and is consequently available to both P and M. This feedback is denoted by two bold arrows leading back to M and P.

In fact, it is a peculiarity of the human mind that the system constantly creates products that are retained and remain available for successive and progressive new reprocessing operations. The 'thoughts that we think' can constitute the basis for modification of the pre-existing functional structure, adding new functions or at any rate modifying the ways in which they are processed. In a word, thought modifies the capacity to think. The existence of this feedback is obvious in the case of certain conscious thought products – for instance, arriving at the solution to a problem, understanding a set of initially disorganized and therefore seemingly meaningless data, or acquiring new (for instance, artistic or scientific) skills, whether or not externalized in a work of art or in some manifestation of the capability acquired. However, the same kind of feedback can be postulated for non-conscious products too, as in the field of artistic and scientific creativity. In psychoanalysis, the analytic process is found to give rise to new functional structures in the analysand that can enable his entire mind system to work in different and better ways. This is the level of typically unconscious productions. Clearly, then, the output of the mind system need on no account possess the character of a manifest, or conscious, phenomenon. Here psychoanalytic science can furnish many illustrations. These also demonstrate the points discussed above, which must be examined in detail to facilitate understanding of our model.

2.5 Primary mental structures

Of the eight points presented for analysis in the previous section (p. 46), the first two fall within the province of psychophysiology. Psychoanalysis has concerned itself mainly with items 6, 7 and 8 – of course from its own standpoint – and to a much lesser extent with items 3, 4 and 5. Yet psychoanalysis may have a contribution to make on these too, and Freud himself had already made an attempt; it is a matter of explaining the genesis of psychic processes in terms of the organization and processing of the data of external reality. The resulting input comprises, first and foremost, the reality of interpersonal relations: the analytic relationship, the focus of attention of all psychoanalytic authors, consists in providing the analysand's mind system with the input supplied by the analyst's mind. This input, though, cannot be understood generically, for it will always consist of the transmission and modulation of physical media, encoded so as to convey meanings, whose decoding by the patient will always depend, first, on processes of sensory reception and, second, on how these are processed by the patient. However, the first order of signifiers conveyed by the analyst's input must be identified in more detail, in terms of the 'external' elements of which they are composed – that is, the elements reaching the patient's processors from the outside. In this sense, the analyst's input constitutes an external reality for the patient – which he then processes internally. In my view, because psychoanalytic theories stress intrapsychic processes and internal reality, they may have discouraged researchers from taking a complete view of the process whereby psychic structures are created, or structured following the experience of a reality that is, after all, external. The notion of libido in Freud's thoroughly endogenist theory depends on the concept of instinct, which, while, on the one hand, suggesting a biological and innate foundation, on the other hand remains totally indeterminate, so that it is now rejected (Imbasciati and Ghilardi 1989; Ghilardi and Imbasciati 1990).

The problem of the empirical, rather than endogenous and innate, genesis of psychic structures[11] remains obscure even in the Kleinian theories. These do not adequately investigate the origin of the internal objects, but accommodate them instead in an apparatus of innate predispositions, referring them back to the Freudian theory in the general and indiscriminate manner

11 The term 'structure' is used in the sense of a functional structure.

characteristic of Klein (Imbasciati 1994: Chapter 9). Bion invokes learning from experience and the transition from 'sensual' to mental data, and hence from external to internal experience, but does not sufficiently analyse the transition from sense data, which are by definition neurophysiological or at most psychophysiological, to mental data, which are psychological and psychoanalytic in nature.

All psychoanalytic theories, in investigating the complexity of internal processes, have tended to portray the processes generated in this complex internality as endogenous, thus making them appear innate. However, the fact that a process is generated internally does not mean that its origin – that is, its genesis – remains confined within the internal system and must therefore implicitly be referred to the innate and the biological, but signifies only that the complexity of the *internal* transformations has been investigated. It is important not to neglect the *external* input, and hence its sensory origin. The reference to the innate is often left implicit, because unconfirmed, but often serves as a plausible label to avoid tackling the thorny problem of how the external input is transformed and processed in such a way that it eventually has effects in internal reality. Some researchers – specifically, psychoanalysts – see the 'quality' of psychic manifestations as so different from that of their neurophysiological and psychophysiological counterparts that they would rather not consider the latter in detail in their attempt to bring a psychoanalytic approach to bear on the examination of how sensory input becomes psychic reality; this results in a de facto generic invocation of the biological and hence of the innate. The innatists' labels often conceal obscurities which the empiricists seek to confront. The dilemma is most acute in investigation of the neonatal or fetal age at which a mind may be said to have come into being, and the terms in which the mind's origins should be considered – that is, biological maturation versus experience, including fetal experience.

If the mind is considered in accordance with the model presented here, which many cognitivists have espoused by analogy with artificial systems (computers), two conditions must be satisfied in order for a program to process data and produce an output. First, the (neural) hardware of the biological apparatus must be of a certain kind – such as the human brain – and, second, the input must be processed by certain functional structures, which serve, and will serve, to process (in a certain way) all subsequent inputs in accordance with certain programs that are structured step by step. The organization of an experience creates a functional structure that is superimposed on its predecessor and modifies it (at the same time usually enriching it), so that the new experiences are processed differently in accordance with the successive processing structures formed. The first – neonatal – experiences of life must be regarded as decisive for this consequentiality, for they determine the creation of structures on whose functions the type of processing to be undergone by subsequent experiences will depend. Hence our emphasis here on the origin of the 'primary' mental structures.

Referring again to Figure 2.1, let us now turn to the first stages of our model, in which the input is processed – that is, the reception system and the encoding system for the received data. As stated earlier, what is received and encoded are not only data leading to perception proper (that is, conscious perception), but any reception, processing and recording of data processed from the inputs in the system M–P. The feedback circuits are relevant here. First, the behavioural response can potentially modify the input: the individual may approach, move away, or select from the stimuli presented by external reality by an appropriate action. The diagram shows a feedback arrow from O_e to I. We must also consider how the M–P system conditions the receptivity of the specific receptors: both incoming (afferent) and outgoing (efferent) nerve fibres have now been demonstrated for almost all receptors, and constitute central afference control mechanisms. The diagram accordingly includes a feedback from P and C_{rm} to R, shown as a broken line. Of more interest to psychologists are the feedbacks in C_{rm}. The M–P system governs the organization and storage of the output from R – that is, the formation of various possible types of memory traces – by modifying the reading code C_{rm}: feedbacks from P and M to C_{rm} are indicated in the diagram by a heavy line.

The paradigm case (although others could be adduced) is perhaps the phenomenon of perception. Here the M–P system determines the various processes leading to perceptual recognition: afferences are distributed, attributed and assembled differently according to sense, those within one and the same sense being combined appropriately (see Section 3.2).

The last aspect of the feedback circuits to be considered is the way the internal output conditions not only the M–P subsystem but also the previous subsystem, C_{rm} (and perhaps also conditions R directly – see the feedback arrows in the diagram). For instance, the creation of internal products in psychoanalytic treatment – of 'thoughts' in the broad sense, insights and new modes of functioning – not only augments the processing structure (M–P) but also modifies the reading structure that organizes reception. Thus, when analysis has given rise to a certain feeling in an analysand, this may enable him to restructure, and improve, his communication to the analyst. This communication is, after all, mediated by the reception and reading of sensible signals. The communication of affects appears mysterious only on account of its complexity – not because it takes place through preternatural channels.

A fundamental requirement of our theory that emerges from the above approach is that the genesis of affects and, at the same time, of the internal objects must be linked to experience, and hence to the processing of traces derived from the reading of the external world. If the internal object is used by an infant to represent the world to himself, it has a representational value that deserves to be studied; in other words, it must be possible to describe the internal objects in terms of representation. This means that we must examine the relationship between the morphology of external reality and the particular internal representational morphology presented by the internal objects; since the internal objects are derived from representations quite unlike any representation that matches the perceptual reality (for normal adults) of any external object, the problem of comparing them with perceptual reality can be solved only if we succeed in breaking both of these entities down into elements that can be found in each with a different organization or composition. This comparison may be likened to analysis of the pictorial elements of an abstract or surrealist painting (such as those by Magritte or Dalí – cf. Imbasciati 1994: Sections 9.3 and 9.7), in which the individual elements of 'reality' are transformed, jumbled together and superimposed in the different order constituted by the internal reality which the painter sought to represent. The comparison between internal objects and external realities is more complicated because not only the visual sense is involved as in the painting example, but also a whole range of sensory inputs, bafflingly mixed up and far more profoundly transformed (Imbasciati 1993b).

The attentive reader will rightly point out that, since the internal objects are formed mainly from elements of internal provenance – that is, from the 'internal outputs' of the M–P system – does this not mean that they stem from elements of non-sensory origin? Close examination of Figure 2.1 shows that these internal elements were constructed, albeit by way of thousands of processing and stratification operations, from elements received from the outside; the problem is that, during the course of life from birth on (and perhaps even in the fetus), these elements have been processed into structures, which have in turn structured new elements, progressively giving rise to more and more complex structures, so that the ultimate products seem to have very little to do with the elements received at the very beginning. Hence a feeling or affect, especially if unconscious, or a thought, particularly in the field of artistic creativity, seems far removed from the 'senses'. Does this mean that it is impossible to seek the primal elements in the internal outputs that have undergone all these stages of processing?

For all their complexity, the most complex elements – 'thoughts' – must always be made up of simpler elements in various combinations that could be called 'signifying units'. Since the brain does not function by means of energy flows, but by the construction of biochemically (and electrically) encoded sets, can the relevant codes be studied not only in neurobiochemical terms but also by the methodology of the psychological sciences, by seeking corresponding or analogous 'engrams'? In other words, might the various internal outputs, however complex, not be supposed to possess representational value? Such a value will take the form of signifying units, of various degrees and on various levels. The internal object is one of the internal outputs (O_i) in our diagram, and perhaps the most elementary O_i of all; it is a signifying unit employed for the internal processes of the earliest stages of mental life, which the child also uses for the purpose of knowing – however inadequately – the external world (Money-Kyrle 1956, 1961, 1965, 1968), and hence to represent that world to himself in some way. This signifying unit is also used by adults, as analysis shows, to 'perceive' not so much simple external objects – this occurs only in actual hallucination – as the most complex interpersonal situations. The individual relates, develops and is modified on the basis of these 'perceptions'; interpersonal relations involve a cognitive process mediated by the internal objects. Hence, the internal object can be regarded as an elementary signifying unit observable in all mental processes; although it is an internal unit produced internally, its origin in primary experience nevertheless refers back to some kind of external reality experienced through the medium of the senses and its subsequent trace-by-trace transformations.

Although it is not easy to link the genesis of the internal objects to experience, understood in terms of the processing of traces, it is an even more complex enterprise to establish a similar link between affects and the results of the processing of traces, as well as to combine the two processes under a single heading rather than placing them in two separate and parallel orders. In so far as the theory of the protomental seeks to identify certain basic operations within the overall context of the processing system constituted by the mind, affect can be defined in terms of the frequency with which the system tends to repeat certain operations. In other words, what we call an affective dynamic is a 'typical' modality of a given individual in a given situation, whereby he processes a certain response; it is a highly intense 'program', which mostly remains internal, on the level denoted in the diagram by O_i.[12] Within this response, which therefore forms part of a corresponding affective dynamic, certain processes that, so to speak, constitute its 'pieces' can be identified: indeed, psychoanalysis has described the various defences as processes that modulate affective dynamics, perhaps transforming them in successive small steps. Defences are thus 'typical' modes of operation of the mind system, which recur in accordance with the predominant structures in a given individual and with that individual's situation; they are therefore operations, or programs, of the system. So we have a system that processes 'pieces' resulting from the encoding, transformation and processing of data originally derived from the senses – that is, traces on various levels – and if, within this conception, we identify ways of describing certain typical operations in terms of traces that act as elementary operational units, then we can define a given affect as a specific typicality of certain operations, performed on elements always made up of some kind of 'trace'.

A further problem is to overcome the traditional dichotomy between affective and cognitive processes and to achieve a unitary view. The distinction

12 I have deliberately used the notation O_i for the internal output because it also suggests 'internal object', on which the search for the elementary signifying unit is concentrated. Figure 2.1 may be compared with a similar, but modified, diagram in my book on medical psychology (Imbasciati 1993a), whereby the 'mind system' is illustrated in the soma too.

between affect and cognition, while to some extent crudely and immediately useful, is, for our present-day scientific culture, the burdensome and mystifying legacy of a much older antithesis, typical of Western medieval philosophy, according to which two orders of phenomena were attributed to two supposedly different 'substances'. In terms of the unification of all psychic processes, the extensive and overall use of the terms 'mind' and 'mental' introduced by Bion can only be welcomed.

Finally, then, the theory of the protomental specifically seeks to embrace and describe all the successive processes of symbolization performed by the mind system. In this cognitivist, experiential context of progressive learning, use will be made of the specific contribution of psychoanalysis – both of psychoanalytic (clinical) methodology and of all existing psychoanalytic concepts and models that can usefully be accommodated within a single framework. This framework must be consistent not only with psychoanalysis but also with everything we now know from the other sciences of the mind. Hence our aim of establishing a 'psychoanalytic cognitivism'. This term may appear paradoxical, for many cognitivists will not recognize 'true' cognitivism in my theory, seeing it as 'irremediably' psychoanalytic, whereas many psychoanalysts will fear that the term 'cognitivism' implies a simplistic reductionism liable to detract from the specificity of psychoanalysis. Yet the term is, I believe, appropriate, indicating as it does the marshalling of different psychological disciplines with a view to outlining a general explanatory theory that will not only accommodate the clinical theories of psychoanalysis but also validate them scientifically.

The theory of the protomental thus constitutes a kind of 'physiology of the mind', which, while psychoanalytic, is at the same time consistent with the criteria of the present-day discipline of psychophysiology.

Chapter 3

The mind as an information-processing system

3.1 Objects, stimuli, afferences and perception

Analysis of the components of a mental process calls for the development of a body of conceptual equipment whereby a series of phenomena can be described. An appropriate technical language is essential if we are to understand the modes of functioning of the mind in relation to the external world and, later, to its own processes. Given that the first mental operations presuppose that something is received and in some way retained, let us set about defining our concepts by first considering the stages of this outside-to-inside storage operation.

The term *external objects*[1] refers to physical entities, such as this book, the things on our desk or in our room, our faces, or the clouds we see scurrying across the sky; their common feature is that they are perceptible and external to the observing subject (they correspond in part to what Koffka [1935] calls the 'geographical environment'). The concept of a *stimulus*, definable as a variation in the energy impinging on our receiving apparatuses, is distinguished from this notion of an 'object'. In the case of the book, the stimulus consists in the fact that, in a given part of the space in front of me, the (electromagnetic) light rays reflected from this object in accordance with certain laws impinge on my eyes: the wavelength of the reflected radiation determines colour, variations in brightness define outlines and shadow areas, and so on. This set of variations in the energy emanating from an object and received by a sense organ thus constitute the stimulation. In the case of the eye, the retina has undergone a modification due to the light reflected from the book and corresponding to the area of space occupied by the book. The impression on the receptor organs will vary in size according to the distance

1 The adjective 'external' evokes the idea of its opposite, and the term 'internal object' immediately suggests the relevant psychoanalytic concept. The polarity of 'external' and 'internal' will be considered later in terms of a possible psychophysiological explanation on information-processing lines, which may also bear out the conceptions of psychoanalysis.

of the object, and will consist of a chemical stimulation determined by the wavelength reflected by the book. The stimulus is therefore the set of variations in physical energy impinging on a receiving apparatus (*distal stimulus*); the corresponding variation arising in the sense organ (*proximal stimulus*) can also be defined in these terms.

According to how the retinal image[2] in the eye has formed on the basis of the stimulation by the object, a specific combination of impulses will pass along the optic nerve and be processed at various neurological stations on their way to the brain. The set of nerve impulses travelling from the receptor organ to the higher nervous pathways – in particular, the brain – is known as an *afference*.[3] However, in order for there to be a *perception*, not only the afference itself (which in any case lacks biological persistence) but also its processing in comparison with previously stored data must be recorded. Perception is not an automatic outcome: a perceptual fact that excludes a prior experience – that is, a prior memorization of data by the mind – is impossible. Perception implies recognition: perceiving a book means recognizing an object in its shape (rectangular), in the colour of its pages, and in its other characteristics, and recognizing it as belonging to a category of similar objects which our prior experience has perceived and memorized as having a similar function and labelled as 'books'. The experience that enables us to perceive a new book object as belonging to the already known category of 'books' is a total experience, which, however, presupposes partial experiences; in our example, we may consider the experience of shape (rectangular shape, or rather that of a low parallelepiped). Now a partial experience of this kind is not fragmentary in nature, since it is complete in itself and has a meaning, and the same can be said of other – such as tactile – experiences. Perception may therefore be seen as the result of a mental operation based on an already recorded input.

The term 'sensation' has been deliberately avoided here because it is equivocal, midway between 'afference' and 'perception' and bound up with the theoretical context of late nineteenth-century scientific psychology (associationism/atomism). 'Sensation' in fact implies something that is experienced subjectively in the same way as perception – a quality lacking in an afference, which is not experienced but known to exist – but is linked to a kind of biunivocal correspondence between the stimulus and what is perceived (the constancy hypothesis). The term 'sensation' proves to be based on the assumption of a natural correspondence whereby man can know the world as it is, by virtue of a natural disposition or of a correspondence between the structures of the world

2 'Image' here refers to a concept in geometrical optics and does not on any account imply the existence of a corresponding perceptual image.

3 The afference is the input and the efference a motor output.

and those of his mind, and not of the fact that he possesses receptor organs that perform functions of storage and operations of selection and comparison, the end result of which is the subjective experience of perception. The structure of the perceptual experience depends not on that of the external objects, but on the product of the interaction between the structures of these objects, the structures of the receptor organs and our previous experience, so that the perceptual result, or, more accurately, the cognitive result, is always due to a certain processing of the 'stimuli'; it does not necessarily correspond to these, and still less to the external objects that caused them.

3.2 Analysis of perception: engrams and representations

Let us consider the perception of an elementary geometrical shape. If I draw a black disc on a white sheet of paper, I have a perceptual result, which is conscious, expressed by 'I see a black disc'. Although this result may appear utterly simple and automatic, it is in fact not so, for the perceptual object is not determined by the input (the afference), which, while necessary, is not sufficient for generating the resulting perception. The input is merely a variation in the trains of nerve impulses emanating from the retina, distributed in accordance with a certain geometry dictated by the optical characteristics of the lens that projected an image on to the retina. In theory, instead of a black disc, I might see a 'white sheet with a hole in it'. In order for the perceptual object 'black disc' to come into being, a number of 'operations' are necessary, which it is worth enumerating here as follows:

1 The first operation is figure-ground organization: the afferences must be 'read' by the mind system, in order for the line of the retinal stimulation gradient to be seen as subjectively belonging – as a margin – to the overall figure constructed on the basis of the afferences from the area of lesser stimulation (the black disc). This reading conforms to complex laws, studied by the Gestalt school, which are the abstract conceptualization of mental operations performed by the system in accordance with its processing of the afference due to the distribution of stimulation gradients over the various retinal zones. Whether these first operations are innate (in the neural structure) or acquired is disputed; many newborn creatures, including human neonates, certainly lack this elementary capacity.
2 For the 'black disc' to be perceived, recognition of the elementary geometrical shape of a circle is necessary; in other words, there must be a prior memory organized in such a way that this geometrical shape can be recognized.
3 In addition, the mind system must be able to separate the afferences organized in accordance with operations 1 and 2 from other, concomitant,

visual afferences. These may be, for example, extrafoveal afferences: the perceptual object remains the disc – that is to say, it is still constructed on the basis of the foveal afferences – while the extrafoveal afferences do not enter into the eventual composition of the disc, but remain organized as (back)ground, in the perception either of the white sheet or of the ground represented by the remainder of the environment; a stimulation that is also received at retinal level emanates from these objects. This 'separate assembly' of afferences calls for a specific mental operation, albeit perhaps an elementary one.

4 Again, the visual afferences must be kept clearly separated from, yet correlated with, those of ocular motricity (proprioceptive afferences and recording of motor efferences): if I move my eyes, I do not see the object change, but know that I have looked in a different direction. An analogous 'calibration' for the motricity of the head and neck is also required. This assembly, too, is a phase in perceptual 'reading' – that is to say, it results from a certain type of mental operation.

5 Furthermore, in composing the perceptual object on the basis of certain afferences, the mind system must keep the visual afferences separate from those derived from other senses – for instance, from the concomitant auditory afferences; otherwise we should have a 'speaking black disc'. They must also be distinguished from other afferences, such as tactile ones; otherwise, if I see the disc while I am sitting down, the object I perceive will be a 'black disc pressing on my buttocks'. So these 'operations' too are necessary.

6 Enteroceptive and coenaesthetic afferences must also be kept separate; if not, if I look at the disc while I have a stomach-ache, I shall see a 'black disc that gives me a stomach-ache'.

7 The mind system must also perform the operation of placing the object constructed as described above in a space outside the self. In other words, integration with a (bodily) inside/outside is necessary.

8 Finally, the system must keep separate the afferences from information of internal origin, that is, feelings, thoughts, images, memories or precursors of these products. These must be distinguished from what will make up the object located in external space. In other words, integration with a psychic inside/outside is necessary if I am not to confuse perception with imagination, or hallucination with reality. For example, I might see an 'angry black disc' or a 'black disc that is also a horse'.

The perceptual act, then, whose simplicity we had taken for granted, is in fact made up of a large number of operations, which are unconscious. Let us try to imagine what we should perceive in their absence. This can be observed experimentally by sensory deprivation or the administration of hallucinogens: the pharmacological effect of LSD, as well as the psychological effect of sensory deprivation, is to disrupt all the mental operations that normally govern and pave the way for the so-called normal perceptual result. A neonate has not yet acquired the capacity for all these operations: like a person under the influence of drugs or suffering from sensory deprivation, he therefore finds himself in a world of totally different perceptions, and may see thoroughly absurd objects. The afferences are composed into sets completely unlike those resulting from the process we call (normal) perception.

A neonate, then, does not perceive, in the usual adult sense of the term 'perception', because his representations do not allow actual, realistic perception: in view of their dissimilarity, in terms of the composition of afferences, to what permits perceptual recognition, they cannot be deemed to constitute perception. What a neonate 'sees' is not the same as adult 'seeing'. Instead, he arguably has an experience comparable to affect in an adult, which he uses actively for orientation in his environment. Is this experience unconscious? The distinction between affect and cognition disappears in the neonate, as does that between conscious and unconscious. Yet, it is precisely with these first phantasmagorical 'units', which are more like affect than perception, that the neonate commences his mental life. What relation do these units bear to psychoanalytic descriptions of the dawn of mental life in a newborn? Money-Kyrle (1968) points out that the primary 'internal objects' (such as the 'breast'), while unlike any representation, are used by the infant to represent the world to himself in some way. Can our description of these units, then, be compared to the psychoanalytic description of the first internal objects? Both may be seen as precursors of the representations proper that will make perception possible; they could be called prerepresentational or protorepresentational *engrams*.

The concept of an engram is therefore introduced here to avoid potential errors resulting from the use of the term 'representation'. As stated earlier, this term has been reformulated for the purposes of this study: a 'representation' need not represent reality as it is, but must only represent it *in some way* for the subject. The term 'representation' may nevertheless mislead, and for this reason I prefer to introduce and use the term 'engram', which I define and distinguish from 'representation' as follows. 'Representation' denotes a concept derived from experimental psychology, even if the term is also used by psychoanalysts, and refers mainly to what the subject experiences consciously in perceptual recognition, memory or imagination. 'Engram', on the other hand, is a wider concept, which includes but does not coincide with a representation; it refers to something that may also be unlike any reality – something primitive, formless and virtually unrepresentable, which is therefore wholly unconscious. Psychoanalysis has used the concept 'representation' in a number of different senses (Imbasciati 1991); the term has been applied, on the subjective level, to something much less readily specifiable than a representation proper, which, moreover, may be wholly unconscious. The engram concept, as adopted here (and as proposed and used previously by me – Imbasciati 1983a), is intended to combine the former psychological concepts (conscious representations, as perceptual representations) and psychoanalytic concepts (unconscious representations), subsuming them within the framework of a psychophysiological inference based on the notion of a memory trace, as the foundation for the 'reading' and use of any mental event, whether produced by external inputs or by internal 'products'. An engram is the entity that operates within the mind system, considered either from the psychological viewpoint or in psychophysiological terms with reference to the neural network, to recognize (that is, to read) a given set of inputs and/or internal products as possessing a sense, or a meaning of some kind, that can be used by the functional network constructed at that particular time on the neural network in such a way as to make it operational. We shall therefore use the term and concept of the 'engram', as defined here, to denote everything recorded in the neural network that makes it a functional network operating as a mental structure, whether on the most primitive level (the first differentiations) or on a more developed plane, such as that of perceptual representation, the formation of affects and the most complex cognitive functions.

The analysis of perception thus leads to a concept – the engram – which can encompass the whole of mental functioning, its origins and development, so that the affective phenomena described by psychoanalysis can be accommodated within the context of cognition. An engram must underlie any function of the mind: the observed clinical entity we call 'affect', whether conscious or inferred, is the epiphenomenon of a functionality laid down in the mental structure. For the mental events labelled affects, the engrams will be more difficult to identify than in the analysis of perception, and will concern compositions ('sets') of memory traces, in which those of external inputs are found to be 'mixed up' with those derived from the system's internal processing. Examples are operation 8 above and operations 6, 7, 8, 8a and 8b below.

For further examples, let us again consider the neonate, or indeed the fetus. A neonate lives in a 'phantasmagoria' that is not affect, not perception, not consciousness and not the unconscious. For a better understanding of this phantasmagoria, let us analyse a more complex perception than the above. A good example is the feeding breast. If an adult were still sucking at the breast, his perception would be along the following lines: 'With my mouth and also my hands, I see and feel a hemispherical, white, soft, warmish object, with a protuberance inside my mouth, from which I can feel a tasty liquid issuing, and I identify this object as having an existence outside myself, as a part of a person who puts herself in a particular relationship with me.' Such an 'object', experienced in this way, is defined as a 'real object' and a 'perceived object'. It is the result of complex mental operations, which can be examined in succession with a view to considering what our experience might be if they were not performed by the mind system, or were performed in a different way.

1 The retinal afferences are grouped together at the stage of central-level decoding into a non-arbitrary 'set', which is not determined by the afferences themselves, but results from processing, or mental operations, whereby a figure distinct from a ground is perceived.

2 This figure is recognized in elementary geometrical shapes – as a spheroid. The memory store is required for this operation.

 a The spheroid is subsequently recognized as a 'breast', by mnemic comparison with images of previously known objects.

3 The afferences organized for the perception of an object of this kind are kept separate from other visual afferences – that is, concomitant, retinal-peripheral afferences.

4 The retinal afferences are grouped together by a precise process of calibration whereby they are distinguished from proprioceptive ocular afferences and from the recording of outgoing efferences; the same applies to the musculature of the neck and trunk. The afferences are calibrated in accordance with the efferences.

5 Non-visual sensory afferences are processed and organized in a particular way, involving the following components:

 a Tactile and buccal proprioceptive afferences are organized similarly to those described for operations 1, 2 and 2a, to identify an 'object' situated inside the mouth; they are also referred to the same perceptual object by way of the organization of visual afferences, but are distinguished from these as a different (tactile, buccal) subjective effect or as a different relationship of the same object with different parts of the subject's own body (the mouth instead of the eyes) – otherwise, the result would be synaesthesia. They are also distinguished from buccal motricity (an object introduced into the mouth is one thing, but my mouth touching and sucking it, or spitting it out, is another; the soft, firm nipple is one thing, while the softness of the buccal mucosa is another). In addition, an anterior motricity (sucking) is distinguished from a posterior motricity (swallowing).

 b The gustatory and olfactory afferences are organized separately and duly coordinated with the other afferences by processes similar to those described for operation 5a. As a result, taste and smell are referred to the same object as has been seen with the eyes and felt with the mouth, but are distinguished by a different mode of relationship of the same object with different functions of our bodily parts, and not as an integral quality of the object.

 c Tactile proprioceptive motor afferences from interactions of the hands with the breast are also organized in precise ways similar to those described for operations 5a and 5b.

 d Concomitant vestibular afferences, or afferences concerning other bodily movements, are also organized in separate sets and at the same time linked together by processes similar to those described above.

e Concomitant auditory afferences are organized along similar lines. If they were not, there would be a 'speaking breast' or, if a car happened to be passing in the road at the time, a 'breast making the noise of a car'.

6 The same applies to bodily afferences, such as stomach-ache, well-being or hunger. In the absence of these operations, a child with a stomach-ache perceives a breast that gives him a stomach-ache; if he has a lesion or wound, however small, he has been hurt by 'that breast'; and if he is hungry, there is a hungry breast.

7 The mind system performs a further processing operation on the afferences, placing the object, constructed as described above, in a space external to the subject's corporeality – external for the part of the breast that is objectively outside and internal for the nipple that is inside, and, moreover, inside a precise part of the body, namely, the mouth. The type of relationship the object has with the subject's own bodily self is identified in this way.

8 Information from within the system (feelings or thoughts), concomitant with the afferences, is kept separate from these, and assembled into a set that forms part of what is defined as the subject's psychic self; in other words, the subject distinguishes himself from the object, and distinguishes what the object causes him to experience and supplies to him from what he thinks or imagines about the object.

a If the subject is about to bite, or imagines swallowing (or if a representation of a motor sequence of swallowing or biting somehow arises inside him), or if a state definable as displeasure, rage or aggression, according to whether certain operations are or are not active, can be identified in his mental system, he will not perceive a breast that can devour him, out of hunger or rage, but will keep the processed inputs from the object assembled differently from those relating to his own psychic products (thoughts or affective states).

b This breast is linked to the whole feeding person, but is also distinguished from her, as well as from the subject himself, and the internal information mentioned with reference to operation 8a undergoes further complex processing in relation to the person, in such a way as to permit the initiation of a dialogue, of which the limited act of sucking is just one element.

The above detailed description will enable us to imagine all the combinations that would be possible if one or more of the operations mentioned were lacking or were performed differently.

Our description in fact illustrates the operations occurring in an adult, or rather, those that would occur if an infant at the breast were to perceive in the same way as an adult. In reality, these operations occur only partially, if at all, in a neonate, and then only after a few weeks. In particular, the later operations in the enumeration do not occur. Broadly, none of the operations are observed at birth; after a few weeks, operations 1 to 4, then 5, and then 6 arise progressively, while the last operations come into being only after the child has been weaned. It is in fact precisely the absence of these operations throughout the first and second years of life that distinguishes the mental (and also perceptual) life of infants; psychoanalysis describes these stages in terms of the paranoid-schizoid and depressive positions.

The above enumeration can be used to help us imagine what might be experienced in the absence of one or more, or all, of these operations – not only in the case of a neonate, but also sometimes in adults under the sway of powerful emotions. It also suggests the existence of corresponding engrams, progressively constructed in the child's mind. Comparisons and analogies are identifiable in the observation of drugged states (LSD) in adults, in certain pathologies, and in states of sensory deprivation.

Another possible comparison is with dreams. Here, too, the boundary between image and representation on the one hand and affective sensation on the other hand is blurred. In my view, this may be because dream representations are not always quasi-perceptual in character. A mind system lacking such representations will be incapable of performing the perceptual process. It can therefore be argued that a neonate does not perceive (in the strict sense of the word), but has experiences more comparable with the adult phenomena of dreams, delusions, hallucinations or, more broadly, affective states – in particular, unconscious affects.

The above concepts and terms, of course, denote unconscious processes; what can be reported by, or observed in, the subject, is their epiphenomenon, from which we infer the – unconscious – processes that have given rise to it. Neonatal observation studies must be considered in the same way. Experimental studies of perception do not tell us that a neonate perceives, in the adult sense of the conscious recognition of real objects that are experienced as real, but do measure behaviours from which laws concerning perception are inferred; these indicate the results of preperceptual operations of which the mind system is capable at successive stages. Only a part of these processes can be inferred. Perception proper is the end result of a complex labour of the mind, which is obviously unconscious, and much of which remains to be investigated (Imbasciati 1986: II, Chapter 1). Can the psychoanalytic approach be used for this purpose? If so, perception could be considered along the same lines as the manifest dream (Imbasciati 1989a).

At any rate, our analysis of the component operations of perception indicates that perception involves a succession of different organizations of the received afferences. It is each of these organized sets, or *engrams*, and not the afferences themselves, that have a representational value. This is so whether this value is considered in terms of the complete end result of the perceptual process as a whole, in which case the representation is said to match reality (the match is not necessarily faithful) – that is, it is a representation proper – or, in the overall process of perceptual construction, in terms of partial organizations of afferences which, although precursors of complete perception, nevertheless constitute representations within the mind. This value too is reflected in perceptions, which we describe as inappropriate or hallucinatory, as in the examples of sensory deprivation and drugged states or, more broadly, of neonatal perception or the perception of a baby in the first months of life. In other words, a complex representational world is inherent in even the simplest of mental operations, which seem simple only because they are automated, as in normal adult perception. This representational world, while manifested in perceptual or quasi-perceptual states, is actively present throughout the operational system of the mind.

3.3 Primary signifiers of the representational world

It is implicit in the concept of the representational world as defined above (which, as we shall see later, can be regarded as homologous to the manifestations observed in clinical psychoanalysis[4] – dreams, small children, etc.) that any incoming (impinging) aggregate of afferences, depending on the type of reading carried out, has its counterpart – that is, is 'recognized' – in a corresponding representation, or rather protorepresentation, which must necessarily be preserved as memory. Otherwise, without memory, we should lack the relevant type of reading – that is, the type of input encoding that occurs at a given present moment. Memory means that every type of representation is made up of sets of traces of inputs previously so assembled in an engram.

These mnemic configurations (the term 'configuration' may be inappropriate, because it does not correspond to the precise 'figures' to which our adult experience has accustomed us) arise with respect not only to inputs of external origin, or from the body, but also to those stemming from and produced by the interior of the mind system – namely, feelings, experiences and thoughts. According to the above analysis, especially of the final stages of the process, the gradual identification of internal states, which are seen to be

4 Some analysts may find the idea of such a homology puzzling, for they are accustomed to deal, and skilled in dealing, with representations or internal states in terms of subjectivity, both of the patient and of themselves (counter-transference). As a result, when we refer to these in 'objective' (or objectifiable) terms, in accordance with our present approach, they may suspect an error involving a 'perversion' of the essence of internal states as manifested in analysis. The theory to be outlined in this book, however, far from denying the value of clinical psychoanalysis, seeks to 'explain' it (Chapter 2), as well as to describe how it deploys its therapeutic effect.

attributed, for example, by neonates to the perception of external objects (what psychoanalysts call 'projective identification'), must necessarily have a specific representation, retained as memory, whether or not this memory is used to 'compose' the strangest and most absurd phantasmagorias, or is instead employed for a more correct recognition of an external or internal reality, or to distinguish between these two realities. These representations have nothing to do with the possible conscious recognition of an emotion, feeling or thought; similarly, representations referable to external reality may enter into perceptual, or rather preperceptual, processes without ultimately leading to a conscious perception. In other words, the perceptual process is one thing, while perception itself is another. Analysis of the process leads us to infer the need for the operation, within the mind, of various representational units (or reading programs, or mnemic sets for the recognition of aggregates of inputs, for which we use the single generic term 'engrams') having thoroughly heterogeneous and diverse 'compositions'. Just as these representational units operate for the encoding (or recognition) of external reality, so, too, they act on internal reality; and this they do on various levels of appropriateness or confusion (hallucination), extending in adults even to the level on which the encoding assumes the subjective character of consciousness – the consciousness of perceiving both external and internal reality. At this point, the subject may even attain to the recognition of his own affective impulses.

The 'true levels of appropriateness' of this recognition can be regarded as falling within a continuum – in particular, within that of the recognition of internal products. At one extreme, these may be read as possessing a meaning totally divorced from their recognition as internal products, so that they are attributed instead to the outside (projection), as if they were external entities (in these cases, a thinking self distinct from the thoughts that are thought often does not exist). Alternatively, the reading they receive may endow them with a certain degree of recognizability as processes produced by the system of an initial rudimentary thinking self. At the other extreme is a reading that leads to an insight – that is, one that entails the appearance of some kind of consciousness. All the intermediate steps along this continuum could be called unconscious experiences. In this context, the mind proves to be a system for processing the data received both from the outside world and from its own internal production; and the resulting elaborations are constantly evolving, in the same way as the whole of memory is continuous transformation. So the mind is revealed as a complex information-processing system, which is thus capable of 'operations' organized in increasingly sophisticated programs over the course of life.

The concept of the representational world may therefore be reflected in the presence and operationality of an infinite number of semantic units, which the mind system can use in succession for 'reading' reality – not only external reality, but also, and in particular, internal reality. The actual operationality of the mind's working can be conceptualized in terms of the processing of these units: each type of operation involves a particular type of processing, and each operation calls for a memory of that operation; the 'functions', which are internal, also require specific representations, preserved as traces to enable them to operate. All these units, which are semantic entities standing for both external and internal reality (including its 'functions' – that is, its character as a process), constitute an infinite progression of *signifiers*, which create and support the various successive signifieds and make them usable. These, considered from a different point of view, constitute mental work – or, as will be shown later, the process of symbolization whereby the mind constructs itself.

The crucial problem in describing such a development of the mind system is to identify how the first operations arise: what is the origin of the first input-processing operations, what programs govern them, and are these programs innate or acquired – and, if the latter, how are they acquired? The dilemma mentioned in the previous chapter (Section 2.2) arises again in this connection.

When a neonate succeeds in unifying within himself the experience of satiety with the sensation of something entering the mouth, an initial mental representation of a feeding entity – a breast or bottle – has already formed, although a perception of the breast, mother or bottle need not be presupposed. For this reason, the investigation of the formation of the mind must focus on how the first afferences can be aggregated in terms of a possible meaning, even if this meaning is not readily definable in terms of sensory representation, and still less of consciousness and rational categorization. When a child has linked together in a single whole the experience of satiety and the tactile experience of something that has entered his mouth, he has formed within himself a unit endowed with meaning that can be called a representation, or protorepresentation, which will be usable for further operations when, a few weeks later, he will face the task of putting them together with visual afferences and perceptually recognizing the bottle or the mother's face.

But when does the baby 'learn' to make connections of this kind?

3.4 Learning and self-construction of the mind system

This study is directed principally toward investigation of the origins of the mind, or *protomental development*, and toward the construction of a corresponding model. In this examination, the innate biological event, which is identical and obligatory for all individuals, is distinguished from the processing-related event constituting the emergence of the 'mental', on the basis of which individuality will subsequently be constituted. The example of the patellar reflex was given earlier (Section 1.4) to analyse an obligatory response to a certain stimulus, in which the variation of mechanical energy gives rise to an afference, which is processed at neural level and immediately translated into a motor event: the bending of the leg proves to be an automatic act which we experience only *a posteriori*. Furthermore, the event is repeated identically whenever the stimulus recurs. A similar problem arises with embryonic development: the manner in which cells, starting with the morula, differentiate and multiply depends on the genetic program contained in the chromosomes, so that all individuals with a particular chromosomal endowment – with that particular program – will develop identically. A different problem, constituting the initial basis for the presupposition of a mental model, arises when an external event – a stimulus – that gives rise to a change in the biological event modifies the structure not only in the here and now when it takes place, but also in the structure's subsequent capacity to react to other stimuli or to the same stimulus – in other words, when a stimulus is somehow recorded by the structure in such a way that the structure's possibility of reacting is modified.

The paradigm of such an event is learning, in the conventional sense of the word: an experience modifies a subject who has that experience, so that his subsequent functioning is modified in relation to that first experience, to its repetition and to any other subsequent experience. However, while such a model appears obvious in learning at mental level, it is also valid for phenomena other than those clearly recognizable as mental, such as simpler events when a given stimulus impinges on a structure and modifies its subsequent functioning. Such a model seems applicable also to biological events. For example, a particular physical or chemical agent in the mother or fetus may modify fetal development in such a way that subsequent interactions with other stimuli are thereby also modified. In this case, the 'experience' has been recorded, causing later experiences to be processed differently. This difference is readily observable at biological level in 'negative' terms, in the case of impaired development of the 'normal' functions of the various apparatuses, and in particular of the neurological apparatus; however, such manifestations may also be postulated as, and in some cases demonstrated to be, mere variations. In other words, even in the absence of highly developed learning and an equally highly developed neurological apparatus, it is possible for a stimulus, however simple, to be recorded, or somehow memorized, at humoral or tissue level, thereby modifying the subject's – or rather the organism's – ability to have subsequent interactions. The term 'learning', in the broad sense, can therefore be legitimately used without the need to assume the existence of a nervous apparatus, and still less of a mental apparatus, as in the case of lower animals.

This model of learning, which is also observable at biological level, thus casts a more complex light on the possibility of demonstrating the 'mental'; in other words, at what point, along the continuum extending from a simple interaction with a univocal and obligatory response to a complex interaction with multiple responses and memorization, can the 'mental' be regarded as coming into being?

To try to answer this question, let us reformulate the foregoing using logical and mathematical notation. Let A be the stimulus and S the structure, defined as the – simple or complex – whole of the biological apparatus for the recording, transmission and processing of the response. If A is superimposed on S to give a result (or response) A_1, the following notation can be used:

$$A = S = A_1$$

This applies to a biological event, such as a simple chemical reaction (for example, hydrochloric acid + sodium hydroxide = sodium chloride), as well as to a more complex situation such as the patellar reflex. Hence:

$$B + S = B_1$$

$$C + S = C_1$$

The essential characteristic is that $B \neq C \neq A$ and that A_1 occurs only after A, B_1 after B and C_1 after C – that is, in a system of obligatory responses. Such a system cannot be described as mental. However, there may also be more complex situations, where stimulus A impinges not simply on S but on a structure that has already stored and memorized an A. The result (or response) in the subject is then different:

$$A + (S + A) = A_2$$

$$A_1 \neq A_2$$

Similarly, where an A impinges on a structure that has already memorized an A and a B, this will give rise in the subject to an event (A_{ab}) different from the above:

$$A + (S + A + B) = A_{ab}$$

$$A_{ab} \neq A_1 \neq A_2$$

Another possibility is that a stimulus A may impinge on an S that has already received a B and then an A instead of an A followed by a B. In this case, the response A_{ba} will be different from A_{ab}:

$$A + (S + B + A) = A_{ba}$$

$$A_{ba} \neq A_{ab}$$

In other words, we have a memorization that takes account of sequences and thus processes the data in a more complex way. A further degree of complexity is also conceivable, in which a more complex stimulus (B + A) impinges on a structure (S + A) and gives rise to an event different from the above:

$$(B + A) + (S + A) = B_x$$

In reality, the stimuli we encounter on a day-to-day basis are never unidimensional but always multidimensional (A + B + C + . . . K): even in the extremely simple example of recognizing a book, there will be a wide variety of different kinds of stimuli, such as shape, colour or three-dimensionality. The various series of stimuli impinge on a structure that has already received a vast number of other stimuli, so that a multiplicity of subjective responses are possible:

$$(A + B + C + E + \ldots K) + S + (A + B + D + I + \ldots n) = A_{x1}, A_{x2}, A_{x3}, A_{xk}$$

according to the combination of stimuli impinging on the structure, as well as another set of responses ($A_{y1}, A_{y2}, A_{y3} \ldots, A_{yk}$) according to the structural organization on which this configuration of stimuli impinges.

In addition, each response in turn constitutes a (new) structure, in which the same event will give rise to different learning operations.

The model thus illustrates, albeit approximately, the infinite possibilities of 'learning'. The enormous quantity and variability of learning operations are further increased if the relevant organism is endowed with a biological apparatus that has become specifically differentiated for the storage of learning operations – that is, with a neural system organized in consequence of certain initial experiences, especially if developed and organized as in the human species. The 'mind system', then, can produce different processing results in accordance not only with the experiential situations that act as stimuli, but also with the functioning of the system itself by virtue of its previous operations; moreover, it can structure itself differently according to how it did so in the past. In other words, the 'mind system' is able to construct itself. Furthermore, the information-processing system represented by the construction of the mind system differs from artificial information-processing systems in a particular respect: the capacity to process and, above all, to 'recognize' internal products, as described in the previous section, allows the system to modify its own functioning retroactively. That, in information-processing terms, is the nature of psychoanalytic work.

However, this 'mental' model does not finally solve the problem of identifying the point at which biochemical and biological reactions and processing, once a certain complexity has been reached, come to be accompanied by the phenomenology that can only be described as 'mental' at more developed levels. In our model, the 'mental', being no longer bound up with consciousness, cannot be contrasted with the 'biological' (as the presence or absence of conscious subjectivity), but constitutes something different, a quite particular entity, which, up to a point, seems describable in biological terms; the problem is to determine when, where and how it arises. It is here that the difficulty of identifying protomental processes lies.

3.5 Learning, representation and the assignment of meaning to afferences

The concept of learning, as described in general terms above, will now be considered with respect to the mind's capability, on receiving afferences from the sensory receptors, of learning the differentiated responses that constitute the recognition of different sets of afferences – that is, the various 'forms' that give rise to the different, successive levels of perceptual organization. In order to perceive, the mind must have organized corresponding representational engrams within itself; it must have learned to assemble the inputs by constructing corresponding representations inside itself. It must have learned a code for assigning meaning to the various sets of afferences.

If we are presented with a white disc, before we can say that we have seen a white disc, the set of afferences issuing from the sense organs following their stimulation by the 'white disc' object must be organized and recognized in a configuration. If they are not, there will only be the fact that a certain zone in the relevant space emits different wavelengths that impinge on different parts of the retina; in order for this fact to 'say' something (at psychological level), these differences in wavelength reflected by the white disc, as opposed to the background, which produce specific configurations of afferences (nerve impulses), must be recognized as a set – that is, as a disc and as white – and connected with prior experiences. For this mental activity to be possible, a system made up of a code for organizing incoming messages, a set of decoding rules and a mode of response must develop. This is obvious in the case of more highly developed mental acts such as verbal and verbalizable thought. For this type of thought, the necessary condition is a type of communication that uses the auditory channel; in other words, the capacity for this kind of thought depends on the functioning of the sensory apparatus represented by the ear and activated by a particular type of energy (sound energy). For example, the word 'mummy', considered in physical terms (as an object), is a variation in the pitch, frequency and timbre of sound waves transmitted through the air and impinging on the tympanum and thence the organ of Corti (a sense organ), from which afferences pass to the brain. For these to be translated into psychic terms, the set of variations in wavelength, amplitude and waveform constituting the word 'mummy' must be organized, perceived as a set and recognized – that is, they must not be confused with other sounds and sound sources. Humans in fact have the capacity to select the wavelengths that form words out of the background noise and to organize them into defined sets, connecting each of them with prior experiences of the sets of sounds corresponding, say, to 'mummy'.[5]

5 Here we may consider the difference between human and mechanical perception: it is harder to follow a tape-recorded conversation with several people talking at once than to distinguish the various voices in the natural, real situation.

A person's mind gradually becomes capable of these organizations: a neonate does not possess this capacity, so that hearing the word 'mummy', or two other meaningless syllables, or the sound of his crib being tapped, has no meaning for him (that is, on the level of the adult communication code; for the neonate himself, it may have some meaning, even if different), because his mind cannot yet isolate certain noises as an integrated set and recognize them. Hence, the code for the organization and recognition of certain acoustic signals has not yet come into being, so that these remain asymbolic messages (or, perhaps, symbols of something very different from what they symbolize for adults) and cannot therefore be used by the mind, either in perception or in memory. A succession of stages before the meaning of words is ultimately grasped has been demonstrated within the communication channel represented by verbal language. By the third or fourth week of life, a baby is able to distinguish the mother's vocalizations and lallations from noises made by other objects; that is to say, he has begun to organize certain afferences into signifying units. The stages prior to the achievement of language (Spitz 1957) can be described in outline as the transition from the distinction between noises and phonemes to that between phonemes and words (Brown, R. 1973; Cromer 1974; Weich 1978). This is not the only communication channel or the only source of engrams (signs) significant for the activity of the mind, although it is probably one of the most archaic, as it is already active in the fetus (Manfredi and Imbasciati 1997). The other sense that will be very important for mental operations is vision, whose organization begins after birth. However, communication through this sensory channel, too, is not the mere reception of stimuli. To return to the example of the white disc, in order for this object to be perceived as a 'white disc', it is not sufficient for a circular area of the retina to emit nerve impulses different from the remaining areas. From the retinal afference to perception, even of simple gestalts, there is a qualitative leap, which in fact constitutes the specificity of the 'mental'.

The same applies to all the other modes of sensory reception. In order for something that can be called 'mind' to begin functioning, reception by the sense organs is not enough; the translation of stimuli into afferences must be processed in such a way that the stimulus is recognized as the 'sign' of an external object. In this case, reception becomes the possibility of assigning meaning to the afference; that is to say, the 'external' can be grasped and identified by a subject in such a way that he can represent and interact with it.

It may be supposed that all the objects on which a newborn baby's eyes fall, and hence the stimuli impinging on his retina, have no meaning for him, so that their effect is comparable with that arising in an adult on encountering something formless that he has never seen before, in which no known shape is recognized. In these situations, however, an adult may recognize colours and distinguish light from shade – that is to say, he may organize the whole into a configuration endowed with some kind of meaning for him, even if he cannot recognize any definite form in it. In a neonate, by contrast, not even this happens, since the set of stimuli from the objects and receptors does not meet with an apparatus equipped to discern any organization in them. For instance, he will be unable to distinguish a white square from other surrounding shapes, or to differentiate the square from the background – that is, to identify the square as an object – but might see it, say, as a hole in his visual space. Electroretinograms reveal the same scotopic and photopic components as those observed in adults; the optic nerve emits nerve impulses, so that the neonate can be said to possess a functioning visual apparatus, but the result is like entering data into a computer without a program: the computer can neither process nor recognize the data. The child must therefore learn to organize these signals in accordance with certain significant sets. For this sensory modality, too, a succession of stages can be identified (Vurpillot 1972; Hutt and Hutt 1973; Bower 1974; McGurk 1974; Schwartz and Day 1979).

One of the first sets to be formed is the figure–ground distinction: certain sets of sensory units are perceived as objects in themselves (such as a sheet of paper) and not as holes in the remaining visual space. These first sets can subsequently be stored and used for further processing in the process of symbolization and discrimination. The presumed characteristic of the neonatal situation is what we call 'confusion' – but nothing is in fact confused, since there is a total lack of signifieds; it is only at a later stage that the child will be capable of recognizing increasingly complex shapes, such as the human face in the fourth or fifth week and, later, a specific human face as different from others. This indicates the progressive construction of a capacity to encode optical signals by an increasingly subtle code, so that, if there was an initial category of 'circles' and 'non-circles', it will later become possible to identify, within the category of 'circles', the set constituted by 'faces', and then, within this group, the face of the mother and later the faces of others. Hence there is a gradual multiplication of signifying units and a gradual increase in the complexity of the code.

However, in order for even a simple code of this kind to come into being, the afference must first assume the significance of an external object – that is, its signifying capacity must include its assembly into a set distinguished from the assemblies identified as parts of the subject's (first bodily and later mental) self, and it must be possible for this set to be accommodated within an 'outside' distinct from the self. Only then can the external object be 'perceived'. It will be recognizable as such because it is represented in the mind by the signal constituted by the relevant set of afferences. The problem now is to determine how this capacity for representation is initially formed, what a protorepresentation of a given external form is in the mind, and whether what constitutes an external object for an adult, with its appropriate representation in the mind, is the same for a child. Alternatively, in a child, might there be representations – or protorepresentations – which, while meaningful for his mind, do not correspond to any 'external object' as commonly so defined? This is the case, as we shall see later, for the first internal objects.

The elementary gestalts have been held to be innate; that is to say, the perceptual organization of the stimuli of which they are composed has been deemed automatic and innate. This notion is based on the observation that all individuals – not only humans but also animals – are already sensitive to these configurations of stimuli in the first weeks of life, and that some species have particular gestalts. However, while this indicates the existence of an innate predisposition of the neurological apparatuses to register gestalts as such, it does not preclude the necessity of an experience – whether neonatal or even prenatal – for making this possible. Studies of imprinting suggest the need for an experience, to modulate the innate predisposition. Likewise, a neonate's recognition responses to certain auditory stimuli (the maternal heartbeat) or olfactory stimuli presuppose fetal experience. In other words, there must be an experience as the foundation for the first 'program' whereby the rudimentary 'mind-computer' can begin to recognize certain data – that is, afferences – which assume meaning only after this first elementary processing operation. Moreover, the need for such experiences has been demonstrated by sensory deprivation experiments that show impairment of the gestaltic organization of perception (Schultz 1965; Reed 1979). Although the predisposition of the neurological apparatus to assign meaning to a certain afference is undoubtedly innate, the process whereby the afference assumes meaning calls for the use of an experience and constitutes one of the first steps in the emergence of the 'mental'. Experiments in which autochthonous factors of perceptual organization are made to conflict with the factor of familiarity or learning, and which demonstrate a perceptual organization contrary to experience (Kanizsa 1968), refer not to the experiencing of an afference whereby it is assigned a meaning independent of experience, but to the experience of a set of afferences to which a meaning is assigned on the basis of something that would be physically impossible in the three-dimensional space of external reality, but is possible in the two-dimensionality of a figure. However, sets of afferences that do not correspond to real objects may also be meaningful, and be the fruit of experience, for the mind. This situation will be discussed in more detail later with reference to the paranoid-schizoid position.

Our study must therefore focus on how and when afferences, or rather sets of afferences, assume in the mind the significance of 'something' that is represented and thus recognized. In adults, this 'something' is the so-called representation of objects of external reality. These are images that have an 'appropriate' correspondence – apart from the normal apperceptive distortions (Imbasciati 1967) – to external 'objects' as they are commonly described; although not faithful representations of external objects as supposed by the classical 'constancy hypothesis' (see Section 1.2), these images can nevertheless be linked to corresponding external objects. But in the neonate there are mental representations, with consequent perceptions, or at least protorepresentations and protoperceptions, of aggregates of afferences endowed with meaning even though this meaning does not necessarily correspond to any object in reality and to any of the perceptions or representations of these objects in the adult mind. They are more like the intermediate products of perception described in our analysis of perception as various levels of assembly and organization of afferences. For these reasons, an afference for a neonate signifies not real objects but 'entities' that are meaningful only in the mind of the neonate himself, through the set of operations performed by this mind. The 'bizarre objects' described by Bion (1967) are a pathological example of mental representations which, while having powerful effects in the dynamics of the mind, do not correspond to any 'object' in the usual sense of the term, so that they are also difficult to describe in words.

Continuing our study of the protomental, let us now investigate the transition from afferences, which are in themselves meaningless, to 'sets' of afferences that represent 'something' that has meaning in the first operations of the mind. This 'meaning' initially lacks any correspondence with real sets that identify external objects. We shall use the term 'engram' to denote these sets, to emphasize that their nature – the transition from a biological afference to the quality of the mental – consists in their being recorded by the mind; that is to say, they are memorized, even if it is not easy to find them in memory, and used as signifying units for the first mental operations.

The above consideration in terms of afferences, external objects and perceptions of external reality is based primarily on visual perception, but is also valid for all the other senses, in various combinations. However, it applies, too, to internal reality: here again, the mind must 'work' with its own self-generated products – that is to say, it must use operational units (engrams) for the assignment of meaning to the results of the mind's processing activity. These engrams, by analogy with the situation described above for external reality, albeit on a more sophisticated level, need not be referable to anything identifiable as an affect, experience, feeling or thought that might be recognizable in an adult mind that is 'up and running', but may be mere 'entities' allowing the system to operate. In intermediate states of perception or the phantasmagorias inferred in relation to the neonate's confusion between external and internal inputs, some engrams, while not corresponding to any real object, nevertheless represent 'something' whereby the mind can begin to operate. In the same way, we must postulate protoengrams that are referable not to any internal product identifiable as imagination, affect or feeling, but to something that nevertheless has to do with the first internal products of the system, which the system itself uses to begin to operate, thereby acquiring the skills needed to organize, distinguish and integrate both what is received from the outside world and its own incipient internal products (see Section 2.3). Of course, even less is known about such protoengrams than what can be reconstructed through the process of perception. They may be connected with the beta-to-alpha transition as conceptualized by Bion (1962, 1970).

Much remains to be investigated in these utterly primal processes, especially if they are to be considered in terms of learning. After all, any differentiation of the system, especially one that initiates a modification of the innate biological basis toward an experience-derived system, can be conceived as learning. For example, even the first differentiations described by psychoanalysis (inside/outside, or self/not self), or what Bion calls the acquisition of the alpha function, ought to be classifiable as learning operations. Just as the system performs learning operations from the reading of external inputs, and restructures itself accordingly – that is, it learns to learn new modes of learning – so, and indeed to a much greater extent, it is logical for the same to apply to the internal products. By processing products, the system 'learns' from them; in other words, by reading these products, it performs learning operations and restructures its own capacity to read them and hence to learn. That is to say, it can learn from its own thought. However, it is much easier to identify the processes of learning from external than from internal reality. And there still remains the problem of where and how the process begins.

3.6 Communication and the experience of learning

The problem of the birth of the mind has been approached on the basis of elementary definitions of learning. If the mind is seen as an information-processing system, the process of learning to act on information seems essential, even as regards the subject's learning to use his own first internal products. After all, learning that relates exclusively to information from outside, unless this information can be 'calibrated' on the basis of information on the operational system's current production – especially regarding its own operation – is impossible, or rather, cannot lead to a result usable for effective interaction with reality. For example, a baby a few months old cannot learn to organize the visual and auditory afferences from a rattle attached to his crib in such a way as to learn that there is a certain object, having a certain shape and colour, above him, unless he is at the same time able to distinguish these afferences from visceral and motor afferences from his own body, as well as from those originating inside his own mind – such as a memory. As we shall see later, if he is unable to distinguish an 'inside' from an 'outside' – a 'me' from a 'not me' – he will be unable to organize those visual afferences effectively; that is to say, to arrive at the conception of the existence of something (the rattle) above his crib. He must therefore learn to process all the information – both that received direct from the outside world and that resulting from his own processing – and to 'calibrate' the various items of information against each other. On a much more complex level, in order for an adult to learn something from a given situation, he must process the information stemming from that situation and integrate it correctly with his own memories, imagination, wishes and feelings. The transference–counter-transference 'structure' represents an even more complex level.

In the learning process, the ways in which sets of information are presented to the apparatus having the task of learning will either facilitate or impede the achievement of a good result: if they reach the apparatus in a form already modulated and organized in a context facilitating the assignment of a meaning, it will be easier for the received data to be processed into a meaningful representation. In other words, incidental learning includes the learning of a communication, or learning 'by' communication – that is to say, the information reaching the apparatus that is to learn has been modulated by another apparatus that has already learned. This is an abstract description of what we know of interpersonal communication and of the fact that a child learns more or less easily according to the mother's (or caregiver's) ability to communicate – that is, to offer the infant signifying units matched to his capacity to acquire them.

With regard to the origins of the mind, let us now consider the communication between mother and neonate (and indeed between expectant mother and fetus) in terms of information processing, or learning.

The first mental activity certainly utilizes engrams derived from various sensory channels that are more primitive than those involved in verbal perception. First and foremost, tactile communication must be considered. In adults, for example, a caress, a handshake or, conversely, a slap clearly represents a communication that goes beyond mere afferences. For instance, if we caress someone and have cold hands, the stimulation we provide is of its nature very different from a caress with warm hands, yet the afference, though different, is organized subjectively in the same way as one due to a stimulus with different thermal characteristics. The mind always 'reads' the relevant gesture and stimuli as a 'caress'; the (mental) meaning is thus superimposed on and constitutes something more than the stimulus. Likewise, we may slap someone so 'lightly' that the tactile stimulation resembles a caress, or, conversely, we may caress a person so roughly that the resulting stimulation is as unpleasant as a slap: the meaning results from the mental recognition, and not from the character of the stimulation.

However, this example entails a highly complex recognition of meaning, which is possible only in an adult. In the case of a neonate, we must consider much more elementary messages (signifieds) and signifiers, similar to those that structure animal communication, but no less important on that account. The tactile channel begins to operate very early in human life and is particularly significant, as it is for all mammals, whereas other animals make little use of the coding of tactile signals. To train a parrot or a crow, one cannot simply stroke it; instead, it must undergo a different programme, involving specific gestures or visual stimuli. For training a dog, on the other hand, tactile communication is of the greatest importance. This is because a mammal, which lives in contact with its mother from birth, develops the capacity to organize tactile afferences into signs endowed with shared meanings. Tactile language, although overlaid with other, more discriminating codes in the human adult, persists – for instance, in the tactile communication of sexual intercourse, which expresses meanings that cannot be translated into words.

The tactile code is fundamental to the survival of a child, because it is the channel through which he enters into communication with those around him: he senses the meaning of, for example, a particular way of being held in the arms of one person rather than another. The child assigns an infinite range of meanings, different on each occasion, to the fact of being held in someone's arms, so that he is able to organize these stimuli – vestibular and tactile afferences – within himself and to process them by a large number of mental operations.[6]

Which channels begin to assume meaning for the organization of recognition codes? Among the first that can be identified are the tactile and preverbal auditory channels: a young child – or indeed a fetus (Manfredi and Imbasciati 1997) – can exhibit differential reactions to different types of noises or sounds, and it is therefore likely that he has developed an initial internal auditory decoding organization. At any rate, the various sensory modalities are assembled in a hierarchical system of interactions between the various codes. For example, although the auditory-verbal code is the most highly developed in humans, the visual code retains its importance, so that a visual language, and hence visual thought, persists in adults. A painter, for instance, thinks in terms of painting – that is, he creates visual forms using engrams that he has processed in a particular way, which differs from that of other people, although the use of recognition through images and thinking in images are universal experiences.

Even before a child learns to think in words, he uses images, or visual symbols. A 1-year-old does not use words, and cannot therefore be said to think verbally, but he undeniably thinks. Again, adults, too, can understand each other through gestures – a means of communication by a visual code that we use much more often than we tend to think (Mehrabian 1971; Argyle 1975). What may be described as an understanding between two people is often constructed and mediated otherwise than through the strictly auditory and verbal channel. The feeling of understanding is made up of gestures, looks and other signs that constitute meaningful communications, albeit with the possibility of error (Tauber and Green 1959). Precisely this possibility of error in the transmission of messages is characteristic of the extensive use of non-biunivocal codes in communication between living beings. Furthermore, this is also true of verbal communication proper: the possibility of interpersonal understanding depends on reduction of the margin of equivocality present in every communication.

Much of our capacity to communicate is mediated by visual channels, even if we are unaccustomed to distinguish the semantic units of this language

6 The organization of neonatal tactile experiences is fundamental to the structuring of adult sexuality both in animals (Schrier, Harlow and Stollnitz 1965) and in man (Imbasciati 1998).

consciously – that is to say, we are unable to say that a look in a certain direction, with the eyes turned one way or another, has emitted a certain message, even if we have actually received and decoded it. The same applies, moreover, to the auditory channel with regard to the tone of voice, which in fact enables us to attune ourselves to another person; this means that we are able to grasp acoustic modulations beyond the level of their organization in verbal semantic units.[7]

To sum up, the possibility of mental operations and the forms they assume are based on the capacity to receive and decode a complex set of data mediated by different sensory channels, interacting systematically and therefore calling for simultaneous processing. For this reason, all communication-based processes of mother–child interaction, classifiable in different groups of behavioural phenomena, must be considered (Camaioni, Volterra and Bates 1976; Freedman and Grand 1977). For instance, a neonate may produce a response (or request) expressed through the cellular metabolism, reflected in a change in some of his cells: many skin conditions are a response to the reception and encoding of certain messages. Disturbances of the gastrointestinal apparatus are other expressions of neonatal communication processes. A newborn baby may exhibit vomiting manifestations beyond typical neonatal regurgitation, or develop gastritis or enteritis. Although some of these conditions may of course have exogenous causes, many can be interpreted as a response by the neonate to the environment (defined in Winnicott's sense of 'maternal care') and hence as a result of mental

7 At this point, some psychoanalysts may be wondering why, given the frequency with which I refer here to representations, I have not yet considered Freud's distinction between 'thing-presentations (*Dingvorstellungen* or *Sachvorstellungen*) and 'word-presentations' (*Wortvorstellungen*), with the relevant definitions and connotations, as discussed in a large number of detailed studies, some of them recent and by Italian authors (e.g. Gori 1991). This problem was addressed in an earlier contribution of mine, on affect and representation (Imbasciati 1991). I would merely point out here that the Freudian distinction is irrelevant to my theoretical framework. I realize that this statement will scandalize some psychoanalysts and that my dismissing the matter in a footnote exposes me to certain charges. Let me say only that Freud's distinction sought, on the basis of the psychoanalytic discoveries then accruing on unconscious representations, to correct the concept of a representation held to be obvious by the sciences of the time, based as it was on contemporary notions of perception in terms of naive realism, biological automatism, the constancy hypothesis (see Sections 1.2, 3.2 and 7.1) or cerebral isomorphism (Köhler 1929). Freud had realized that this concept of a (re)presentation was inappropriate and therefore attempted to correct it, but was unable to refute it; its eventual invalidation had to wait for later research in the field of perceptology. Freud's distinction seeks to identify some intermediate levels along the representational continuum portrayed in this book in Figure 10.1; it succeeds in the aim of demonstrating the existence of unconscious representations that cannot by any means be accommodated within the classical concept of a conscious representation, but since this concept is retained as a reference term, the distinction proves to be adultistic and fails to account for what may occur in the neonatal mind.

processing. Some authors (Winnicott 1958; Spitz 1965; Kreisler, Fain and Soulé 1974; Bertolini, Geitlinger, Guareschi and Cazzullo 1978; Gaddini 1978) consider a component of this kind to be involved in most so-called neonatal disorders.

The decisive factor is held to lie in the psychological environment constituted and created by maternal care – an environment which is therefore defined in the sense of information and, in particular, communication. If a child finds himself in an environment unsuited to his incipient mental structure, he may exhibit serious pathologies, which may be not only somatic but also psychic; for instance, his early psychic manifestations, such as response to a smile, overall motricity, motor skills, and discriminatory and perceptual capacities, may be retarded. Alternatively, the baby may develop forms of anorexia, refusing food (breast or bottle) although manifestly hungry; this situation may result in severe malnutrition. Again, the child may exhibit restlessness, manifested in continual crying for no apparent reason. Some children cry when they are hungry or cold or when they are bathed; this means that they have learned to use crying to obtain something – that is to say, they have learned to decode stimuli from the environment, as well as from their own bodies, to assign meaning to them and to organize them in such a way as to produce a meaningful result. A child who cries when he is hungry and seems to be calling his mother shows that he is able not only to perceive the unpleasantness of an organic state of hunger, but also to connect it with the possibility of calling the mother – in more abstract terms, he connects it with the memory of an experience of need satisfaction and hence with the possibility of summoning it back. Crying, used as a signal with meaning, implies that certain mental operations having a specific logical sense have been performed.

Crying can be used to signal any other kind of need – as when an infant, after being held in the mother's arms and 'cuddled', is put down suddenly. In this situation, crying can assume the clear meaning of a demand for the 'cuddling' not to end so suddenly but instead to be prolonged. From such behaviour, we may infer the formation of a link between a pleasant sensation, perceptions inherent in the mother figure and the recognition of a state of discomfort: mental processing of this kind is connected with the possibility of causing the source of this gratification to reappear through the crying-related demand. Conversely, in some situations, the organization of mother–child communication by crying may meet with difficulties, owing mostly to inadequacies on the part of the mother in decoding the first messages and helping her child to learn to use them. In this case, crying may assume disorganized forms comparable to schizophrenic communication, or there may be psychomotor (mental) retardation, sometimes presaging more specific pathologies such as autism.

These observations suggest that crying is a very important signal in a child's psychic development (Winnicott 1957). Where used as a clear signal with a concrete reason, it means that appropriate and correct mental operations have been performed, but if it is so disorganized that its meaning cannot readily be understood, this indicates that inappropriate, unsuccessful operations have been carried out – operations that could be called 'erroneous', in that they have not led to 'solutions' to the problem of communication.

For this reason, if the mind is seen as an operational system implanted on the neural system, which it modifies according to experience, its construction can be described in terms of learning operations, and explained in terms of memory traces that undergo constant restructuring. From this point of view, the construction of the system appears to progress by a series of 'operations' definable as 'correct' rather than 'incorrect' according to whether they lead, more or less directly and quickly, or conversely do not lead, to solutions that give rise to mental structures effective for interacting with reality. The various processes identified by psychoanalysis in the psychic development of children, and perhaps also those of fetal development, can therefore potentially be redescribed in these terms.

Of course, whether the signal is used clearly to indicate successful, reality-discriminating operations or in a confused way that betrays insufficiently discriminatory mental processing, also depends on the receiver – that is, the person who decodes the signal (the mother). The reference to the 'clarity' of a message, after all, implies the relationship between its sender and its receiver; hence, not only must the child learn to use different types of crying or to cry in situations where the signal conveys a comprehensible meaning, but also the mother must be able to decode the signal on the basis of her discriminatory capacity, whose subtlety will be based on her own childhood experiences and the codes she learned when she too was small. The capacity to understand a neonate's messages is not conscious but unconscious – indeed, mothers are often unable to verbalize what their children are expressing, even if they actually respond appropriately. Winnicott (1965a) refers in this connection to the 'good enough mother', which means that the maternal codes whereby the child's communications are decoded have been constituted with sufficient clarity. A mother who has a 'good' relationship with her child, enabling him to develop in the best possible way, is able to understand when, how and why he cries, and can in turn help him to learn to cry meaningfully. The child has 'learned' because the mother has modulated an input for him that is consistent with his learning capacity.

Hence, the child's ability to organize, as quickly as possible, 'correct' mental operations – that is, ones allowing increasing synthesis of perceptual and sensory elements – which can be transformed into memories that are in turn transformed into images, fantasies and, later, thoughts, depends on the mother–child relationship and on the possibility of establishing a shared communication code (Freedman and Grand 1977) whereby learning can take place. The mother must for this purpose possess a structure capable of communication without excessive mistakes, so that the child receives non-contradictory communications; he will then be able to distinguish between these communications and to produce a comprehensible response that can in turn be used by the mother as feedback on her child's mental operations.

If this communication code is established incompletely, defectively or, still worse, contradictorily, greater or lesser dysfunctions will be observed in the mother–child relationship: the mother will not succeed in establishing appropriate and subtle communication, and the child will in turn be unable to develop effective, non-contradictory and communicable mental operations. To simplify with the example of crying, if the mother fails to understand what the child wants, she will not know what to do, and will proceed by trial and error; the repetition of fruitless attempts will not provide the child with the feedback necessary to enable him to structure his crying. The more disorganized the mother's attempts to make the child stop crying, the less able he will be to assign meaning to his situation of discomfort and hence to respond in such a way as to evoke appropriate action by the mother. In this case, the contradictory information and the inappropriate mental operations that fail to lead to a satisfactory solution are multiplied many times over. The result may be a vicious circle: the less able the mother is to decode the child's messages, the less she will be capable of supplying him with clear communications. The child will then receive ambiguous or contradictory messages, making him even less able to undertake appropriate encoding and increasing the level of confusion (but not of meaningfulness) in both himself and the mother. This will give rise to a type of communication experienced by the mother as annoying and irritating and by the child as anxiety, rage, desperation and, later, confusion.

Disturbances in mother–neonate communication sometimes impress an indelible stamp on the development of the child's mind: the deep roots of all forms of psychosis and of many mental retardation syndromes must be sought in problems with the establishment of the first mental operations in the first year of life, based on the relationship between mother and child. Disturbances of the mother–neonate relationship may also be reflected in somatic manifestations – not only in organic syndromes due to abnormal neonatal mental development, but also in the disposition to pathological somatic communication, that is, to psychosomatic disorders.

In conclusion, this chapter discusses psychic structures and activities as an open system that uses the sensory apparatuses and processes information by a series of codes within a communicational context represented by the mother–child relationship, which modulates the progression of learning operations whereby the operational system called 'mind' constructs itself. Using the concepts developed in the first part of this book, we shall now analyse the genesis and gradual constitution of mental processes from a macrogenetic and microgenetic developmental standpoint on the basis of experimental and clinical evidence.

Chapter 4

The engram: afferences, mental operations and internal objects

4.1 The fetal mind and perinatal learning

Let us return to the problem of innate versus acquired characteristics. Does the fact that two children of the same mother and father differ in their characters and other respects depend only on differences in communication – that is, on the differing use of different experiences – or is it due to factors inherent (innate) in the specific individuals concerned, bound up with their individual biological structure as determined by their chromosomal endowment?

It is difficult to determine when one of these factors outweighs the other. Common parlance may be misleading, when children are said to be different because they 'have different characters'. The concept of 'character', used in this way, is ambiguous, because, on the one hand, it refers to something innate and, on the other hand, it implies a process of development ('character formation'), which signifies the use of experiences. Strictly speaking, the term 'character' should be applied only to the latter aspect, although of course the experience used in the formation of character structures is that of early infancy, or perhaps even of fetal life, so that these structures resist modification by subsequent learning. The term 'character' thus denotes an entity that is already the stabilized product, laid down at an early stage, of various environmental interactions, and not an innate factor.

This is not to say that innate factors are negligible. Usually, however, they tend to be overestimated: it is forgotten that 'innate' has the specific meaning of 'biological' – that is to say, bound up with the chromosomal structures of the individual, which determine the biological development of his organism, including brain development. Yet brain development, in which there might (although this is not yet certain) be individual chromosomally determined differences, should not be confused with the functional development of the brain – that is to say, with the fact that a *particular* biological apparatus begins to record *certain* experiences, which form the nuclei of *certain* mental operations, or of a *certain* operationality, that in turn give rise to a *particular* use of a *given* experience and will constitute an interaction with a *certain* type of communication. This is indicative of an individual development connected with experience, even if presumably determined by an initial, innate biological structure. This functional development gives rise to a corresponding development of brain structure: experiences and learning operations determine corresponding organizations of the neural connections. Cerebral morphology itself – especially synaptic proliferation – is determined by learning. In animal experiments, cortical morphology is found to depend on the learning which the animal has undergone. For this reason, besides the traditional notion that neural structure determines learning, it must be remembered that, conversely, learning also conditions neural structure. 'Functional development' is inseparable from morphological development: the functional structure that constructs itself by successive learning operations is therefore also a neural structure. The concept of 'character' mentioned above, together with its 'structure', although it can be regarded as laid down in the brain, is a learning-related acquisition.

The significance of innate factors is usually overestimated owing to the difficulty of observing experiential factors at the very early ages to which they belong. It is forgotten that the overvaluation of innate factors is tantamount to emphasizing the importance of individual chromosomal endowment. There may well be innate anatomical differences in cerebral structure as between individual A and individual B – although no experimental evidence of this has hitherto been forthcoming – but, for the time being, the theory that attributes the difference to functional structures, constituted by the products of the first mental learning operations, appears eminently satisfactory. 'Structure' can usefully be likened to a computer: two computers with identical hardware configurations will perform different operations according to the first data stored and used for programming (software).

In this connection, it is worth recalling that the separation between biological structures, which are transmitted genetically, and acquired factors is not clear-cut. Mancia (1980) points out that the protein synthesis constituting the substrate of a process of periodic genetic programming takes place during rapid eye movement (REM) sleep, in which the previous day's experiences are reprocessed in dreams together with mental products of internal origin. Referring to the work of Jouvet (1978), Mancia (ibid.: 315, translated) writes: 'REM sleep and dreams may be regarded as also providing the ideal framework for genetic programming of the mental functions on which dream thought is based.' The resynthesis of RNA is linked to that of DNA, thus permitting the transmission of what is acquired (that is, 'learned', in the broad sense, considering that, as discussed earlier, learning means not transferring a content from outside to inside, but processing); hence the concept of dream intelligence (Imbasciati 1983a).

The problem of innate versus acquired characteristics is complicated still further when the fetal mind is considered. The existence of a fetal mind, and hence of experience, perception, learning, processing and communication (between the expectant mother and the fetus) – from the sixth month of gestation on has now been demonstrated; the relevant research is reviewed in Imbasciati (1996, 1997), Manfredi, Imbasciati and Ghilardi (1997), Manfredi and Imbasciati (1997), and Della Vedova and Imbasciati (1998). Again, even before the sixth month, it is impossible to rule out the existence of a mental function of some kind, or of learning in the broad sense (see Section 3.4) – that is, the possibility that different stimuli might give rise to subsequent differences in the capacity to process them and that there is a form of learning not mediated by neural structures.

Hence, the specificity of a given form of mental functioning must depend on differences in experience, understood as the personal history that commences with the first environmental interactions, including intrauterine interactions, and the first communications. 'Differences in character' relate not to the individual's biological structure but to its functioning. This thesis is confirmed by studies of monozygotic twins (having the same chromosomal endowment) and dizygotic twins (who are biologically different). Comparison of the two types of twins may enable us to determine the relative roles of genetics and of experiential factors in the causation of character differences. Although there are fewer psychological than biological twin studies, considerable psychological differences in both monozygotic and dizygotic twins have been observed. It therefore seems legitimate to conclude that biological determinants have a relative effect on psychological differences.

A theory to explain the beginnings of mental activity must necessarily start from the fetal mind. In the fetus, the first experiences give rise to a form of processing greatly determined by the type of mother–fetus communication, accompanied by the relevant learning, which leads to the first operational capacities of a nascent information-processing apparatus. This apparatus *will learn different modes of learning*, and this fact will determine the entire subsequent course of development. The first stored programs will determine the type of programs constituted at every subsequent instant, and so on for every experience, from which the capacities of the mental system will be drawn. However, our present knowledge of fetal psychic life is scant, incomplete and partly hypothetical, although the research began a good 20 years ago and has now given rise to a substantial body of literature; it is a totally inadequate basis for proposing a theory of the construction of the mind. On the other hand, the senses 'opened up' at birth, together with the known data on perinatal life, derived from experiment, from direct observation of neonates and infants, and from psychoanalysis, appear fundamental to the information-processing conception presented in this book. For this reason, the model to be developed below is based mainly on what we know of neonatal life; some of the various fragmentary hypotheses that may emerge from knowledge of the fetal mind receive merely incidental mention, in the hope that a fuller treatment will become possible at some future date, when more is known about the subject.

Previously inactive apparatuses commence functioning with the act of birth, so that the neonate's mind receives a large quantity of new and unprecedented afferences, which it must learn to know. It is therefore confronted with the task of organizing these afferences and of transforming them into meaningful sets, engrams and perceptions – that is, into elements that can be used for some kind of mental operation.

The fact that the eyes commence functioning does not mean that the baby perceives his surroundings, but only that the eyes begin to emit the relevant afferences. These can be used for mental operations only to the extent that they are processed in such a way as to assume a meaning that is preserved beyond the period of presence of the afference (that is, transmuted into memory). In other words, the afferences must be processed in such a way that sets of afferences are identified as signifiers of signifieds to be stored in memory; that is to say, the afferences must be 'read' by some kind of apparatus that can 'recognize' them – an apparatus possessing engrams appropriate for recognizing the sets of afferences as having a meaning of some kind. A newborn baby does not see objects in the same way as adults do; everything reaching him is meaningless. For an adult, seeing means recognizing – recognizing shapes, colours and objects. None of this takes place in a baby's mind when his eyes begin to function; he must learn to receive the afferences, to organize them meaningfully and to store their organized traces. The more the mother helps him to establish clear, non-contradictory communication without confusion (decodable two-way communications), the more appropriately and the earlier this process will take place.

The eyes of some animals are closed in the first days of life, and some human babies keep theirs shut too; this is a way of allowing contact with this new afference to arise gradually. The initial impact of the visual afference is presumably bewildering and stressful – as if an adult were suddenly to acquire a sixth sense having the intensity and continuity typical of visual afferences. As adults, we should no doubt feel invaded and bombarded by something we had not yet learned to organize; the neonate's closing of his eyes thus signifies that he is trying to limit and control the impact of this new invasion of stimuli. This is consistent with the advice that newborn babies should be kept in semi-darkness to help them become accustomed to visual afferences gradually, thereby facilitating the organization of these afferences; Winnicott (1957) says that the mother should present the world to them in small doses.

A neonate receives not only visual but also thermal afferences. Following the constant temperature of fetal life, he suddenly experiences at birth a lower temperature; the resulting substantial thermal excursion can be defined as a sensation of cold. The phrase 'sensation of cold' denotes something that, in the adult mind, has undergone a certain organization, concerning the graduation of differential quantities of heat that have come into contact with the various parts of the body, and in particular the capacity to recognize this sensation. All this is yet to come in a neonate, who presumably feels invaded by an afference stemming from the entire surface of his body in the form of something unpleasantly unknown.

The third apparatus that begins to function – and this is vital to the constitution of the 'container' (Bick 1968; Meltzer, Bremner, Hoxter, Weddell and Wittenberg 1975) that will allow appropriate psychophysical development of the neonate – is the tactile sense. In the amniotic sac, the fetus is more or less sensitive to pressure – that is, to the application of overall or localized pressure due to the mother's changes of position, movements and touching. However, immersion in the amniotic fluid eliminates or substantially reduces tactile sensitivity; the sensitivity that does exist is due mostly to intermittent contact between the skin and the inner wall of the amniotic sac. Hence tactile sensitivity may be said to be fully activated only after birth, when the baby begins to be touched, handled and dressed. All this, too, must be organized in accordance with knowledge of external reality. An adult may recognize an object by touch, but in a neonate this capacity is still to come. After all, while different afferences are occasioned by a touch on the right hand as opposed to the left foot (because the impulse passes to certain brain cells rather than others), in order for this to assume a psychic character, it must be transformed into perception, with recognition of the touched part. This entails organization and comparison of the traces of the various afferences.

The apparatuses that send gustatory and olfactory afferences to the brain have functioned at reduced level before birth, responding merely to biochemical variations in the amniotic fluid. At birth, however, the receptors are exposed to stimuli of a quantity and quality that call for much greater discrimination, so that the baby must organize the afferences into something internal in order for them to be usable mentally. We are unaccustomed to regarding olfactory afferences as the foundation of our mental operations, which we place on a more developed level, basing them on codes borrowed from many other senses – specifically, sight and hearing – with reference to the learning of language (Vincenzi and Imbasciati 1993). Yet olfactory and gustatory messages quite probably make up a large part of neonatal experience; and what makes them even more important is that they are closely bound up with the new mode of feeding.

Feeding also involves visceral afferences, whose level was previously low, but which are now embodied in an experience called 'hunger' that was completely unknown to the fetus: the resulting discomfort will force the baby to feed. There are innate behavioural schemes, such as the sucking reflex, whereby stimulation of the buccal mucosa gives rise, through pre-established neurological channels, to an organized contraction of certain muscles and hence to the act of sucking. These schemes, however, although necessary, are not sufficient for the neonate to commence feeding and to continue to feed regularly. A use of experience must be organized on top of the innate reflexes, so that the baby learns to know hunger and to feed: difficulty in feeding is a common vicissitude of neonatal life.

Besides these experiences of sensory afferences, a neonate has one last order of fundamental experiences – namely, respiration. He must now have recourse to an apparatus never previously used, and this new experience also probably has traumatic connotations. Before commencing regular respiration, the baby experiences a few moments of insufficient breathing or none at all – that is, of partial, physiological asphyxia at the moment of birth. He must learn to use the enteroception of hypoxia to initiate the automatic operation of the respiratory muscles so that air enters the lungs, thus putting an end to an internal state of discomfort; the commencement of this new activity is also a stressful experience.

This entire set of afferences, which are inevitably described in adult language, constitute the raw data that must be organized into meaningful form for the baby, at however simple a level; in other words, they are the elements used to perform the first mental operations, which constitute the basal experiential structure. A specific type of processing may arise in accordance with the particularity of each sensory channel; that is to say, engrams with different developmental vicissitudes will be constructed. That is the basis of my earlier description (Imbasciati 1983b) of the 'symbolization chains' specific to each sense (see also Chapter 8).

4.2 Traces of afferences, engrams and internal objects

One of the first mental operations a neonate must learn to perform is the 'reading' of afferences. When he has acquired the capacity to obtain an appropriate result from the 'complete' reading process as described in Section 3.2, perception proper will have been attained; that is, there will be a sufficient match between the external object and its representation. This constitutes an engram, whereby perception, or recognition, of the representation becomes possible. However, in all preperceptual states and even when perception has been achieved, much later, during the first 2 years of life, inputs of external origin continue to alternate, and be mixed and confused, with bodily inputs and those stemming from internal mental products. In this case, the representations do not stand for anything existing in external reality, or else they do so in a manner described as hallucinatory, delusional or persecutory, if metabolized in the paranoid-schizoid mode (see Chapter 6). Yet it is with this first type of representations that the neonatal mind learns to operate – and also learns a kind of knowledge, which we qualify as hallucinatory or delusional, of the world. The neonate's internal 'images' – or, better, engrams – do not correspond to external objects: his *internal objects* do not correspond to those of reality, made up as they are of traces of sets of afferences assembled in ways inconsistent with the generation of representations appropriate for perception proper. Moreover, the product of an external input is confused with data from inside the incipient functional system itself. Such engrams correspond to the entities on which psychoanalysis has traditionally concentrated.

The term 'internal object' immediately suggests the corresponding psychoanalytic concept, with all its 'history' (Imbasciati 1993b). For this reason, the context in which I have used the term so far, and in which I shall continue to use it below, may be found puzzling. Some psychoanalysts may be alarmed at the 'denaturing' of a psychoanalytic concept, or the loss of its specific psychoanalytic sense, and perhaps even imagine that I am unaware of its proper meaning. I hope that a close reading of this and the subsequent chapters of my book will show them that the internal object to which I refer is the same as that featuring in the traditional psychoanalytic literature, but approached from a different standpoint. Psychoanalysis describes it in substantially affective terms, whereas mine are protocognitive. The internal objects have a representational quality for the infant (Imbasciati 1991), and therefore perform a cognitive function. Might it not be his internal objects that enable the baby in some way to apperceive reality and know the world in his own way? This function was illustrated a good four decades ago by Money-Kyrle (1961, 1968). Moreover, since psychoanalysis has discovered that the internal objects remain active in later life, modulating adult thought and behaviour, it ought to be possible to assign them a cognitive function for adults too. That is the tenor of our description. Our account seeks to define the internal objects not only as affective objects, but as mnemic sets – sets of traces of afferences assembled into representational units which, while not representing any object existing in reality, are nevertheless essential operational units of the mind. This approach is specifically intended to furnish an 'explanation' of the origin of the internal objects, as well as to describe that origin in cognitive terms comparable with the relational terms of the psychoanalytic tradition (Imbasciati 1991, 1993b, 1994).

Freud's oeuvre is pervaded by a twofold aim: to afford an *understanding* of psychic processes for the analyst and the analysand, and to offer a scientific *explanation* of these processes to psychoanalysts and other scientists, consistent not only with the *description* given to facilitate understanding, but also with the findings of other sciences (see Sections 2.1 and 2.2). The first aspect has constituted the basis of clinical psychoanalysis and the psychoanalytic method, while the second underlies the theory – specifically, the energy-and-drive theory – by which Freud sought not only to understand and describe but also to 'explain' psychic processes. The energy-and-drive theory was able to 'explain' psychic events because it was consistent with the relevant findings of other sciences at the time of Freud. The corresponding sciences today refer to the mind in terms no longer of energy but of information processing (Bucci 1997), so that Freud's energy-and-drive theory no longer has explanatory value. In the theory developed in this book, I offer a possible explanatory key to psychic processes on which present-day psychoanalysis can agree with other contemporary sciences, based on information processing rather than energetics. A rudimentary attempt along the same lines was made by Peterfreund and Franceschini (1973). In the context of a theory possessing explanatory value, the term 'internal object' is used here with a view to tracing its origins in the psychophysiology that underlies mental processes.

The internal object has been invoked with reference to the particular engrams that permit appropriate reading of sets of sensory afferences – that is, those that permit perception, in the sense of the perception of external objects as 'objects' existing in the external world, separate from the subject, independent of his activity and having characteristics of their own. This reading can occur only if the subject possesses corresponding engrams. However, in the achievement of this organization – that is, in arriving at the construction of *this type* of internal object – how are other inputs (the confused, mixed-up inputs that constituted the first internal objects) assembled? The bodily inputs may reasonably be supposed to have their traces organized in engrams allowing the achievement of adequate proprioception and coenaesthesia. But for what products within the system? What are the possible vicissitudes of memories, or rather protomemories, traces of motor afferences and, in particular, the mixed-up, confused assemblies (external and internal together) of the type described in Section 3.2? Are they destroyed, or do they persist in deep operational structures?

Psychoanalysis has described internal objects with precisely these characteristics of confusion between self and object (that is, narcissism) and between 'inside' and 'outside', and of the intermingling of corporeality, mental products and external perceptions relatively remote from reality and subject to a multiplicity of variations and vicissitudes – such as paranoid-schizoid rather then depressive metabolism, 'part objects' rather than 'whole objects', and so on. The deep affective structures, especially the most primitive, are described in similar terms. How can all these descriptions be explained? What is their origin? In particular, what is their neurophysiological – or, better, psychophysiological – basis?

From the point of view of psychophysiological functions, the internal object of psychoanalysis, or rather the various constantly changing and developing types of internal objects, must necessarily be connected with engrams, made up of the assembly of complex, mixed-up memory traces of the kind inferable in the neonate or deducible from analysis of perceptual organization (see Section 3.2).

The psychoanalytic concept of an internal object can now be linked to the sense of the term put forward above, so that it can be used in an extended sense to emphasize its cognitive value in relation to its psychophysiological support, the engram. The psychoanalytic internal object can be seen as a particular engram; and having regard to the concept of the engram, we propose to use the term 'internal object' in an extended sense, which includes the internal objects that have been described by psychoanalysis. These have a complexity of their own; for example, the internal object is referred to as something separate from the self, but, on the other hand, more primitive pre-object states are also adduced. The perceptual internal object has aspects of separateness, albeit referred to external reality, similar to those described for certain types of, or rather moments in, the vicissitudes of the psychoanalytic internal objects. Therefore, the perceptual object can perhaps be seen, in cognitive terms referred to the outside world, as a precursor of the maturation later to be undergone by the internal objects that govern the deep affective structures – that is, that function, having regard to the relevant engrams, as 'readers' of internal reality. Conversely, the initial engrams of the stages of perceptual maturation are found to be analogous to the most archaic psychic structures described by psychoanalysis – that is, those relating to so-called pre-object states.

In our usage, then, the notion 'internal object', corresponding to 'engram', subsumes the psychoanalytic internal objects, considered in cognitive terms. However, not all engrams correspond to the internal objects of psychoanalysis.

The purpose of all the foregoing propositions is to arrive at a better description and explanation of how the origins of the mind – actually the mind system – are to be sought in experience. That is the general aim of the present 'theory of the protomental'. It seeks to explain how the mind can come into being from experience and, at the same time, how it is that the quality of its birth and subsequent development is determined not by the environmental elements in themselves – the sensory stimuli as physical events impinging on the receptors – but by the ways in which these are processed and by the outcome of this processing in the constitution (construction) of certain functions. In other words, while experience is indeed connected with the environment of the subject-to-be, it does not consist in that environment; this experience is first and foremost that of the *specific* individual – internal experience, we should call it, if the term 'internal' did not suggest a fully formed subject, rather than the individual *in statu nascendi* under consideration here. In the case of such an individual, it is experience derived from the senses that must be considered first, for it is impossible for a neural system with zero input to develop. At the same time, however, we must take account of the way in which the afference will be processed at the various levels of integration and compared with ('calibrated' against, as we put it earlier) what begins to be produced at the very start by, or in, the nascent system, which will constitute the development of the mental structures. From this point of view, experience, in its origins, is sensoriality integrated with the primitive nucleus of eventual internal activity. Experience thus proves to be, at one and the same time, experience of the outside world and internal experience.

Bion's 'learning from experience' can be accommodated within this framework and assigned a meaning that remains psychoanalytic, as intended by Bion and as conceived by psychoanalysts (experience as internal experience from, or through, which 'learning' does indeed take place – that is, the capacity to think develops), while at the same time the greatest possible emphasis, as regards the origins of the mind in a young baby, can be laid on external reality. Internal reality is constructed on the model of the metabolization of its external counterpart at the beginning of life.

Psychoanalysis has always concerned itself with internal reality, but by describing it in affective terms it has made for an endogenist conception of affects (cf. Freud), while their experiential origins have tended to be overlooked. As shown above, these internal objects can be seen, with regard to their origins, as cognitive nuclei, and their genesis can be related to the processing of external reality. They can now be considered from both of these viewpoints. Put differently, the affective aspect is nothing but a summary definition of the group of mental processes whose cognitive quality is less clear – not only in young children but also in adults, in whom so-called affects are the foundation, which is obscure to introspection but emerges clearly from clinical investigation, of all cognition.

Classical psychoanalysis has described the internal object in terms of affects, but that is an epiphenomenal account, using an adultistic analogy, of what are in fact the first modalities of knowledge. The affects are basic cognitive schemata (cf. Plutchik 1980). The distinction between affectivity and cognition is a legacy of Western philosophy and of pre-1930 psychology. The word 'affective' was applied to the form of mental activity that lacks the biunivocal precision and verifiability observed in conscious, rational cognition as expressed by verbal thought. To repay our debt to Bion, to his conception of 'thought' and to his 'grid', we must take the investigation of the origins of *learning from experience* to its limit – by integrating psychoanalytic knowledge with the findings that have accrued in the last 20 years in the neighbouring psychological and neurological sciences. Now that the instinctual, energy-related, endogenist approach to psychic processes espoused by Freud, in line with the other sciences of his time, has been abandoned as untenable, these processes must be explained differently. 'Object'-based psychoanalytic theories place the origins of psychic structures in the relational experience of the earliest stages of life; however, besides the *how*, which is indeed relational, more attention needs to be devoted to the *why* of these origins in experience. That is the aim of the present theory of the protomental. Besides identifying the context – which is indeed that of relationship – in which the construction of the primary psychic functions take place, we must describe in more detail, and also 'explain', how and also why what is presented by the senses is transformed into the mental domain. In other words, the description of *how*, arrived at on the basis of clinical practice, must be supplemented by an explanatory hypothesis on a psychophysiological foundation consistent with the principles of the present-day neurosciences.

It would be naive to consider that perceptual objects can be constructed on the basis solely of afferences, for an afference in itself does not include the element of memory: the neurological input is quickly dissipated unless it is 'read' in accordance with prior memorized processing operations; the mere recording of afferences does not suffice for the result to be used for mental operations. The level observed in mental operations after the second year of life requires afferences to be organized within the mind in sets definable as images, or rather perceptions, endowed with meaning in the sense of correspondence with an external object of some kind. Appropriate knowledge is possible if a correspondence can be assumed between something inside ourselves and external objects. However, achievement of the operational mental level observed after the second year of life is necessarily preceded by a complex and prolonged process (which is precisely the subject of this book) whereby the mind constructs these 'images' – that is, establishes an appropriate correspondence between external and internal objects. Prior to the achievement of this correspondence, on which the mental operations observed after the second year of life depend, the mind processes afferences by organizing them in *engrams* that serve for more elementary operations; these, while manifestly inappropriate for the purposes of knowledge, nevertheless constitute the first attempts at, and the first steps in, more appropriate operations, so that they themselves have a cognitive quality, even if they have hitherto been described by psychology in terms of feelings and by psychoanalysis as internal objects and fantasies.[1]

In fact, the first differentiations, or origins of functions, regarded by psychoanalysis as pre-object and pre-symbolic states – such as those described by Bollas (1992) as states of being, by Bleger (1967) as agglutinated states or by Greenspan (1997) as deficits in symbolization – can also be seen as engrams.

The study of the protomental, or of the very earliest cognitive processes, thus focuses on all the intermediate phases of organization of afferences, as well as of the inputs from the incipient activity of the mind system itself, which, all mixed together, go to make up the first engrams. The process of development can now be seen as one of progressive interlinking of engrams and of constant differentiation between them, and between the stimuli of external reality that must be organized to make up perceived objects and those derived from incipient internal reality, culminating in the construction of what psychoanalysis calls fantasy (or 'phantasy' in the Kleinian sense), internal objects, affects, feelings and, ultimately, mental representations proper, defined as the biunivocal correspondence between something internal and something external. In the course of this development, one of the first

1 One of the difficulties many psychoanalysts have in conceiving of the representational quality of internal objects, and hence with the use of this term in relation to sets of memory traces, is due to the implicit assumption that perception is a simple and automatic process, which automatically gives rise to traces that are consequent upon and faithfully represent what has been perceived. Owing to this assumption, any representation-related term or concept is taken as signifying a faithful and appropriate representation; conversely, anything not thus automatic, distinct and faithful is ascribed to so-called affectivity.

differentiations to be considered is that between 'external' and 'internal', and the progressive reduction of the primal confusion between them.

4.3 Perception, memory, imagination and hallucination

To attain a knowledge deemed to 'match' reality, the mind system must have developed sufficiently to distinguish between 'inside itself' and the outside world. A fundamental characteristic of adult mental activity (or of that of a child over 3 years old) is surely the distinction made by the mind system itself between something referred to external reality – that is, something perceived or thought of as existing – on the one hand, and, on the other hand, something felt to belong exclusively to the mind; that is, the distinction between objects experienced as external and 'objects' experienced as products of our own imagination, thoughts, memories or daydreams – that is, *our own internal states*, or, in a word, internal objects (see Sections 2.5, 3.5 and 4.2), at the various levels of their subsequent progressive differentiation from each other. This distinction of external from internal objects allows the construction of more subtle distinctions between individual internal objects – that is, objects referable to a representation permitting a perceptual recognition of some kind, rather than objects experienced as inherent in memories, or ones felt to be 'imagined' but nevertheless referred to a hypothetical 'outside', and finally objects experienced as hallucinatory in nature.

Five different primary levels of knowledge of reality can thus be identified:

1 An 'internal' level – something rudimentarily felt to be 'inside', which is at first exclusively bodily.
2 Perception, which implies something internal that is, however, experienced with reference to something existing in external reality, present to the observer in perceptual recognition.
3 Memory, which, on the other hand, refers to an internal engram having the same characteristics as the percept, but deemed to be 'not currently present', which is therefore linked to an initial concept of temporality.
4 Imagination – that is, internal objects that do not correspond to those of present external reality, but which could possess such a correspondence; the non-correspondence with the present is experienced by the subject.
5 Hallucination: here the internal objects do not correspond to external objects but are nevertheless experienced in a confused way as if they did.

Note that all these levels characterize a 'knowledge' totally independent of the dimension of consciousness that it might *possibly* assume.

What underlies these mental objects and the distinction between them is the way the mind itself works, with its various recognitions, nexuses, links and comparisons – that is, 'thoughts'. The working of the mind, too, is an 'object'

– that is, something that happens on a more or less recurrent basis or exists as a capacity in the mind itself: something which, although proper to the mind, does not assume the so-called objective characteristics of the other four 'objects' described above, even though, strictly speaking, it, too, must be regarded as an engram that comes to be laid down, or constructed, in the mind.

A normal adult is able to distinguish appropriately between various levels of his internal reality and hence to acquire an appropriate knowledge of external reality. A neonate, on the other hand, must learn to distinguish between percepts, memories and imagination, and therefore cannot have an appropriate knowledge of external reality before achieving this internal degree of discrimination; he therefore hallucinates. Taking the categories of an already developed system as our starting point, let us now proceed backward with a view to inferring how they are formed and in what terms the working of the mind can be described when the distinction has not yet been attained.

The *perceived internal object* is an internal entity – an organized whole endowed with some kind of meaning or, if you will, an engram, experienced in relation to the outside world and, within limits, matching the external object that gave rise to the afferences organized as the relevant perceived internal object. This 'match'[2] (which characterizes an engram that may be regarded as a 'representation' proper) allows the subject to attribute the category of 'truth' to the percept – that is, to base the possibility of knowing the truth of external reality on perception. Comparison of the percept with the other kinds of internal objects constitutes the basis of the capacity for distinguishing between 'true' and 'false'.

The *remembered internal object* signifies that the mind contains an image, matching an external object and experienced as internal, relating to the past presence of a corresponding external object. In this case too, the distinguishing of memory as such permits the experience of a judgement of truth concerning the past existence of the external object that gave rise to the persistence of the memory. Corresponding to this 'judgement of truth' is the capacity to distinguish the *imagined internal object* from those discussed above – that is, the capacity to experience an internal engram endowed with meaning as internal and originating in the mind without any correspondence to an external object that exists or has existed in reality. In this case, the 'truth' of what is imagined is distinguished, as internal truth, from the falsity of a supposed or supposable presence of a corresponding external object.

Finally, the mind's capacity to make primary judgements of truth or falsity

2 There is in fact no such thing as a perfect match; perception always involves some degree, however small, of distortion of reality (Imbasciati 1967).

is based on the possibility of distinguishing the above three classes of objects: the presence of hallucinated internal objects – engrams endowed with meaning and experienced as referring to external objects even if they do not correspond to any objectively present external object – involves the existence of an external object falsely believed to be real. Where these objects are present, the subject is unable to distinguish them from those of the first class – percepts – or, consequently, from objects of the third and second classes, as all of the first three classes are confused; it is then impossible to achieve the capacity to make a true/false judgement based on a knowledge of reality, or to arrive at what will later be the experience of time. Conversely, the absence of this category entails the possibility of distinguishing between the other three and an appropriate knowledge of the world of external objects, achieved through their representation in internal objects. The hallucinated object therefore constitutes a category of internal objects on whose absence from the mind the presence, discrimination and operationality of the others depend; hence, it is a negative category in the functioning mind. Its appearance as a positive element is a dysfunction of the mind. Hallucination, while a peculiarity of abnormal minds, is present in the neonate and persists until his mind gradually becomes capable of distinguishing between perception, memory and imagination and of ridding itself of its hallucinated products.

The four classes of internal objects can be envisaged with reference to the diagram. In this diagram, P, (H), M and I denote the four classes – namely, perception, hallucination, memory and imagination. H is shown in parentheses because this category is only potentially present.

The diagram illustrates the logic of the situation only, and refers to polarities that can be demonstrated when the mind has achieved a quasi-adult level

		External reality	
		Present	Absent
Internal reality experienced as	external	P	(H)
	internal	M	I

of functioning. The above categories therefore do not possess absolute value, for countless intermediate engrams exist in the mind, probably far outnumbering those assignable to the four categories and constituting the 'wealth' of the psychic world. Perhaps the real problem is to describe how the mind can attain the three polarities of functioning evinced in conscious thought while at the same time using all the intermediate engrams.

In discussing perception, we pointed out earlier that its characteristic feature is the 'match' between the percept and the real object. This match is always relative, owing to apperceptive distortion (Imbasciati 1967, 1978a); strictly speaking, no perception is ever 'perfect' – that is to say, a faithful reproduction of the external object – for all perceptions contain within themselves a modicum of distortion, or superimposed memory (the apperceptive mass), and imagination-cum-projection, so that a modest degree of hallucination is inevitable. Yet the four categories retain their distinct functions, which should be conceived in probabilistic rather than qualitative terms, for the purposes of mental operations. For this reason, all references to a match between P and reality should be understood as signifying a match 'sufficient' for the process P to be distinguished from the other three, thus making possible a number of mental operations adequate for an operational knowledge of the external world – the form of knowledge that characterizes a 'normal' adult mind.

Again, the analysis of artistic creativity, of interpersonal communication, of dreams and of all the manifestations revealed by the psychoanalytic method shows that the adult mind, while capable of conceiving the three categories of conscious thought, works essentially with intermediate engrams. The mind of an infant, and indeed also of an older child, functions in the same way. This is exemplified by Winnicott's concept of the transitional object. For our present purposes, having examined the end products that allow rational thought, we must now turn to the problem of how the mind comes into being in the course of infant development – that is, to the study of all the intermediate steps in mental operations.

A normal adult is able to distinguish appropriately between the four categories described above and thereby to discriminate between three levels of reality, knowing when and how external reality corresponds, has corresponded, or does not correspond, to internal reality. This is the basis of the functioning of the adult mind; if any one of the four categories is confused with others, the result is an *incorrect mental operation* (see Section 3.6) – that is, one unsuitable for a knowledge of objective reality. Yet this *is* possible when the mind is, or rather becomes, capable of not attributing to the outside world an internal object that does not correspond, and has never corresponded, to anything in the outside world – that is, when it is capable of experiencing as 'imagined' what is only inside itself. It takes about a year for a neonate to achieve the capacity to make these distinctions. We must therefore investigate how incorrect operations can ultimately lead to *appropriate operations*, and, in particular, how the former continue to make the latter possible and to constitute the basis of the mind's internal, affective wealth. While this concerns the distinction between true and false, real and unreal, and existing and imagined, it also involves the use of the latter categories for the former, as well as, ultimately, 'play', which endows the former with their 'substance'.

The foundation of mental development is the laborious differentiation between *inside* and *outside*. There is, on the one hand, what is felt to arise in an 'inside', which is initially bodily, the nucleus of an incipient self capable of producing something; and, on the other hand, what exists 'outside' – that is, what the mind has assembled not arbitrarily but in such a way as to permit the recognition of real objects that are actually present. Out of this construction (perceptual engrams) are formed engrams that contain within themselves the experience of memories – that is, something that is outside even if no longer present – and finally of something that might realistically exist in an 'outside'. The ability to process the 'inside' proves to be extremely complex, and develops in parallel with and subsequently to its 'outside' counterpart. After all, a very young child has no experience of a mental 'inside'; an initial 'inside' is exclusively bodily, while a mental 'inside' comes later. Furthermore, for much of the first 2 years of life, 'inside' and 'outside' are liable to become mixed up with each other again and to overlap, and the same applies to a bodily 'inside' and a mental 'inside', as well as to all of these with the various categories of the 'outside'. Typical examples are the paranoid-schizoid situation and modulations of the psyche-soma.

The distinction illustrated in the diagram given earlier in this section entails the correct attribution of something 'inside' firstly to the outside (perception), secondly to the inside (imagination) and thirdly to the inside but with the experience of an external correlation – a 'past' experience as in the case of memory. If what is inside is not distinguished according to the two experiences of 'outside' and 'inside', or is not correctly distinguished (that is, is confused), P coincides with M and H with I. In this case, it is no longer possible to distinguish between the columns of the diagram, between present and absent external reality, because I merges with M, since there is now no possibility, in an internal experience felt to be such, of discriminating between elements experienced as previously correlated with 'outside'. It is therefore impossible for the mind to judge whether an object does or does not actually exist in external reality, or to judge the truth or falsity of an object that appears in the mind but in respect of which it is impossible to know whether it does or does not exist in reality. Everything is perceived, remembered, imagined and hallucinated at one and the same time.

The mind, then, before attaining the capacity to distinguish between inside and outside, may be said to live in a continuous state of hallucination. This is physiological for a neonate. However, while it may be useful to say that a neonate lives in a continuous state of hallucination – because it suggests an adult experience that is understandable, if pathological – this conception is not, strictly speaking, correct, since the term 'hallucination' is meaningful precisely when opposed to imagination and perception. If these do not yet exist, neither can true hallucination. An adult subject to pathological hallucination may refer his internal images to the outside world because he has known, and knows, perception; for him, the hallucinated image is experienced as if it were perceived. But for someone who has never been able to know perception, the hallucinatory state cannot be referred either to the outside or to the inside, precisely because the distinction between outside and inside is meaningless for him. Hence the experience of a neonate, who as yet lacks any experience of discriminating between inside and outside, is not definable in terms of hallucination proper, or of perception, or indeed of feeling; it is a confused dimension, very different from those typical of the adult mind, and therefore not readily understandable. The internal world of a neonate is thus perhaps even harder to understand than that of a psychotic.

4.4 Internal objects and bodily experience

Thinking of a feeding bottle, we have an image correlated with perceptions that match the objective reality of that physical object: what we have 'in our minds' – the engram of the bottle or the internal object 'bottle' – is a definite object that refers to but is distinct from a 'thing' in external reality. A small child does not yet possess this correspondence between internal representations and real objects, nor does he have any consciousness of the distinction between the two. For a baby, the first engram of the bottle will be made up of something to do with the traces of the buccal-tactile-gustatory stimulation probably associated with visceral afferences (satiety or hunger) and also connected with others, of a vestibular and proprioceptive nature, because he is taken into the mother's arms and moved during feeding. In particular, however, even when the child puts together these sets of afferences and retains them as signifiers having some kind of meaning, he is unable to distinguish whether this is something that happens to him automatically within his mind or something that has appeared at a certain point in external reality and is giving rise to the stimuli his mind is attempting to put together; in other words, he cannot tell whether the event belongs to external reality or is internal.

The child's mind thus contains 'sets' that begin to assume an initial rudimentary meaning, but which are not yet either specific or specifiable; these first objects, which are of a quite peculiar nature, may appear extremely bizarre to the adult mind, made up as they are of traces of afferences aggregated in totally unaccustomed and inappropriate ways. It is only later that they will be distinguished and reassembled differently.

Our image of the bottle consists essentially of visual afferences, with proprioceptive afferences that contribute to the assignment of solidity to the engram; for a neonate, by contrast, the proto-image is composed of tactile, buccal, thermal, gustatory-olfactory and, in particular, visceral inputs. These by no means correspond to the objective bottle but instead, in accordance with the distinction made by an adult mind, to experiences associated with its use. For this reason, these first internal objects must undergo an entire process of transformation into engrams that afford a better match with reality and that distinguish various aspects of experience from reality. In the above example, the visceral afferences must be completely separated and detached from the protoengram of the bottle, while being supplemented by visual and manual proprioceptive afferences; buccal afferences must also be 'detached' from the proto-image of the bottle and attributed to the baby's own mouth. In this way, it will become possible to distinguish an external object that is perceived – and moreover one that is perceived in a way resembling the real external object – from perceptions assigned to the child's own body in the action of using that object. The first internal objects will thus progressively be transformed into symbols more appropriate for assigning meaning to external reality, and hence more suitable for more effective mental operations. This will be possible to the extent that sets are formed in the mind which can biunivocally – without the possibility of error – assign meaning to the objects of external reality, to their interactions both with each other and with the subject, and to events internal to the subject. In the neonate, there is as yet no biunivocality between the sets formed in his mind and 'objects' in the strict, objective sense of the word; total confusion therefore reigns, so that it is wholly possible for no distinction to be made between internal objects, external objects, individual external objects or individual internal objects, as well as between external 'objects' proper and entities that are functions, interactions or uses of those 'objects'.

Psychoanalysis has emphasized how all adult mental processes bear a relationship to their first infantile counterparts, which it has described in terms of primary relationships and the structuring of affects. However, in the wake of Freud's endogenist approach (the so-called doctrine of instincts), it was, in my view, implicitly assumed that relationships were structured by affects; and when doubt began to be cast on the endogenist approach, a sufficient explanation was never given of how it was, if affects were not prior to relationships, that *relationships* structured affects. The Master's 'doctrine' (Imbasciati 1994), or the 'Witch metapsychology' (Freud 1937c, 1985c [1887–1904]) – that is, the psychoanalytic mythology represented by metapsychology (Fabozzi and Ortu 1996) – may be said to have met its nemesis here. If the affective structures cannot reliably be deemed primary and innate because organic in origin, the problem arises of how and why relationships structure affects. Here we must not only examine *how* – that is, the relational modalities – but also investigate *why* this experience, since all experience necessarily begins with the senses, is processed in such a way as to produce the primary mental structures we call affects. *That* is the subject of this book:

what are the mental – actually protomental – operations that organize primary experience? Internal objects have long been a matter of debate, but it seems to me that their genesis as the (mental) organizers of data received from the body's external and internal sensory apparatuses has not been sufficiently clarified (Imbasciati 1991).

The fact that the first internal objects are organized by means of the confused mental operations we call affective processes does not imply that these operations are qualitatively different from adult so-called intellective processes. The former, too, serve for knowing the world; it is just that this first knowledge is absolutely inappropriate. Thus, the cognitive processes, in the strict, traditional sense of the term, do not appear suddenly, out of the blue – or purely out of biological maturation, as is sometimes maintained – as if they were qualitatively different from, and on a higher level than, their affective counterparts, allowing the mind to take the qualitative leap from the world of affects to that of knowledge. The cognitive processes in fact develop slowly through the transformation of the first processes, which some prefer to describe as affective – that is, through the transformation of the first internal objects into symbols more suitable for representing reality, whether that reality is external or internal.

This transformation does not imply that the primitive processes are superseded by other, more developed ones: the latter in fact exist side by side with and are superimposed on the former, which remain active. All 'affective' dynamics persist in the adult – or, better, all mental, or protomental, operations remain active in the unconscious functioning of thought. There is no conscious thought without a simultaneous, underlying unconscious thought. The latter is made up of the functioning of all mental operations, ranging from the most primitive to the most highly developed: the conscious aspect of thought is the end product, in which only the most highly developed mental operations – those sometimes restrictively called 'cognitive' – appear. However, all the processes based on more primitive codes persist at an unconscious level.

We must therefore analyse the process of transformation of the first protomental dynamics into mental operations increasingly appropriate for assigning meaning to and knowing reality; that is to say, we must study in detail the process whereby the first internal objects are broken down into more appropriate symbols and codes and into increasingly complex mental operations with a minimum margin of error. The data afforded by psychoanalytic science (Bernfeld 1925; Klein 1932, 1957, 1961, 1973; Erikson 1950; Winnicott 1958, 1965a, b, 1971; Bion 1962; A. Freud 1965; Spitz 1965; Bowlby 1969, 1973, 1980; Fornari 1971; Mahler, Pine and Bergman 1975; Greenspan 1997) will therefore be put together here with the findings of research on the psychology of perception (Bruner 1968; Arnheim 1970; Hutt and Hutt 1973; Bower 1974; Foss 1974; Carterette and Friedman 1975, 1976, 1978a–e; Vurpillot 1972). Our aim is to clarify the transition from what cannot be described as mental (that is, the input) to something that we can begin to call mental, or protomental – something tantamount to a first 'operation' of the mind. Let us therefore set about describing in more detail the formation of the first internal objects, or indeed of the very first internal object.

In the cognitive model described here, the starting point, from which the neurobiological hardware of the mind system obtains the information to be processed, stored and structured in such a way as to form successive functional structures, is experience mediated by the senses – that is, the afference, 'processed' on the various levels that we shall attempt to describe. In this processing, consideration must be given to the starting level – not so much within a single sense or a single order of receptors, as in sets of different sensory inputs, gathered from different receptors according to the prevailing type of bodily experience, which will thus involve some receptors rather than others. In the case of a neonate, attention will be directed to the parts of the body most involved, and hence to the type of afferences to which such involvement, or exercise, leads. The type of bodily experience thus determines the first form of input modulation.

This view is already implicit in Freud's description of the three libidinal phases and the importance assigned to the respective bodily zones. However, it was left to later authors, including in particular Erik Erikson (1950), to develop these notions into a model of the structuring of the psyche by bodily experience (rather than the other way round, the structuring of certain bodily experiences being held to be required by psychic development, as Freud seems to be saying in certain passages). Erikson suggests a reversal of Freud's causality: for him, instead of libido cathecting the organ, thus giving rise to a certain mental structure, it is the exercise of an organ that generates typical structures. In this way, oral, and later anal, exercise generates 'organ modes', which are reflected in modes of general psychic functioning extending beyond the use of the relevant bodily zones. In this connection, Erikson assigns great importance to the exercise of motricity, thus suggesting that the parts of the body involved in infantile development are by no means confined to the erotogenic zones on which Freud focused. Sullivan adopts a similar position (Imbasciati 1994: Section 12.1).

In our predominantly empiricist (rather than innatist) approach to the origin of the first functional structures of the mind, we shall consider all the bodily zones or activities that prove to take priority in neonatal life (first survival, and later relationality). We shall thus consider the activity of sucking and the oral zone – not because we believe that these zones are activated by some kind of (libidinal) energy or innate predisposition (apart from the sucking reflex), but, more simply, because a major part of neonatal activity actually involves the mouth; it therefore follows that all the afferences entailed by this activity assume a central role in the construction of the first schemes of mental functioning. Similarly, it follows that all the concomitant afferences – not necessarily from a single sense and not necessarily of external origin – enter into the first modulation on the basis of which the first 'mental' sets (the first retainable traces, the first functional 'programs', the first 'structures', in accordance with the general outline put forward here) will be processed. We shall seek the transition from the sensory to the mental level in the first modulation carried out by virtue of the fact that the afferences are combined in a certain set owing to the exercise of a specific bodily activity. Here memory traces will be constituted, not so much from the first protorepresentations of objects as from the first operational functions acquired by the system. 'Memory traces of functional structures' means that the acquisition of any new functional mode involves a structure retained in memory as a specific 'program' for the relevant functioning.

Since the prevalent exercise of one bodily zone rather than another is by no means negligible for the beginning of the functional schemes of the mind, examination of the set of afferences making up the input to the processor from the experiences of this exercise will involve an analysis of the relevant mental 'modes' of functioning. These, and any of their products that are retained (mental contents), are therefore likely to bear the stamp of the bodily zone involved in the primal situation. In other words, the symbolizations derived from that experience, as well as the respective structures that will perform the symbolopoietic function, may be characterized by their zonal origin. We may thus speak of oral symbolization to denote – as we in fact intend – the type of symbolization and the manner of symbolizing (that is, the characteristic of a specific function) that stem from the mental structuring of the experience mediated by the exercise of the oral zone. Anal symbolization can be viewed in the same way. However, attention will also be directed, if Erikson's principle is further developed, to other functional apparatuses. Thus, we may invoke not only muscular or locomotor symbolization, as already discussed by Erikson, but also auditory, visual, olfactory, gustatory, respiratory and vestibular symbolization, as well as masticatory symbolization, prehensorial symbolization, and so on (Imbasciati 1983b).

The problem is to analyse the type of influence exercised by a certain zone in the process of mentalization (or symbolization) and the sequence of the various types of processes; after all, there is no doubt that a specific type of prior structuring following a certain bodily experience materially affects the ensuing type of symbolic process. For instance, if a particular form of symbolization is attributed to oral experience, this means that certain functions will be structured from this experience, and that the relevant type of functioning will determine the processing, and the structuring in subsequent functional structures, of afferences stemming from a prehensorial, anal or urethral type of experience. So we must, in effect, establish a chronology for the priority of exercise of the various bodily zones or the various functions of specific senses. In this context, the theory of the protomental seeks to describe the origin and formation of the first engrams and of the corresponding internal objects, as well as their representational value for the first cognitive functions. The starting point, in the present state of our knowledge, is taken to be oral experience. However, it should be noted that doubt has lately been cast on this traditional chronology by studies of fetal psychic life, which suggest instead that other – for example, auditory – experiences may come first. Insufficient information on this possibility is yet available, so that we must, for the time being, base our theory on the traditional starting point of oral experience.

4.5 The 'breast'

The traditional view of the internal object called 'breast' is that of a set of memory traces resulting from the experience of sucking; however, this entity does not correspond to the organized set denoted by the phrase 'representation of a breast or feeding bottle' (as entailed by adult perception), or to the other set that corresponds, again in adults, to the representation of an act of sucking. Instead, it corresponds to a different form of processing of the traces of afferences inherent in the overall experience (including the relationship with the feeding adult), all mixed up in a way quite different from the combination underlying perceptual representations.

Let us now reconsider the analysis of perception given in Section 3.2 in the light of the foregoing, re-examining how traces of afferences are organized in the experience of adult perception of a breast or nipple with a view to understanding how different a baby's experience of the 'breast' may be.

1 The tactile and proprioceptive buccal afferences from a teat (or nipple) in the mouth are grouped together in accordance with the elementary level of processing known as the figure–ground organization. This comprises the division of afferences from a given sense into two groups, one of which is read as belonging to and forming the object, and the other the (back)ground. If I see a black disc in a white field, this means that some of the retinal afferences are organized in such a way as to make up the 'black disc', while others form a ground. If this were not the case, I might see a white screen with a round hole in it, or a different, no more readily identifiable object. The figure–ground organization (and its reversal) has been thoroughly studied by experimental psychology, and not only for visual perceptions. In the tactile-proprioceptive perception of our example, only some of the buccal-tactile-proprioceptive afferences are integrated in such a way as to constitute the object identified in the mouth; others are used to make up the 'ground' of perceptions of the buccal cavity containing the teat. Again, both are 'calibrated' differently according to the motoricity used to contain the teat (muscular proprioceptive

afferences and recording of motor efferences).[3] If this figure–ground organization does not take place, the teat cannot be distinguished from the mouth that contains it. That is probably what happens in a neonate: the internal object 'breast' does not correspond to any representation whereby the object 'nipple' might be distinguished from the baby's own mouth.

2 The traces of afferences that identify as an object the nipple or teat contained in the mouth must include a recognition of this object in its elementary forms – that is, a certain three-dimensional geometrical form with a certain consistency and deformability. For this purpose, the memory store must be drawn upon. If it is not, a protorepresentation not containing this comparison will not allow any recognition of something identified as an object.

3 The recognition must also include a precise comparison with similar experiences of that object, in order for the entity identified as an object having a certain form and consistency to be subsequently recognized as a bottle. If it does not, a protorepresentation not involving this 'memory' will not allow the object to be recognized when present. This can readily occur in a very young child: instead of serving for recognition of a breast present in external reality, the internal object 'breast' often leads to the perception of a different object, which may be persecutory or 'bad'.

4 The buccal-tactile-proprioceptive afferences are kept separate from other, concomitant afferences of the same type, from other bodily zones, whether near or distant – such as those resulting from the baby's being held in the mother's arms. If they are not, a protorepresentation lacking this separation will lead to the perception of a 'breast that holds in its arms'; and that is indeed what the baby represents to itself through the internal object 'breast'. A similar separation is necessary for proprioceptive-motor integration, as well as for the other senses – for instance, separation from the vestibular traces that usually accompany feeding. If this is not the case, a protorepresentation without this distinction will lead to the representation of a 'cuddling breast' – as may indeed arise in a baby with regard to the internal object 'breast'. Similar distinctions, necessary for the adult representation of a nipple to arise, apply to other senses: gustatory, olfactory and thermal afferences must be distinguished as stemming from the precise object concerned rather than from other, concomitant sources. The adult representation of the nipple identifies smells, tastes and warmth as belonging to the particular object rather than stemming from other sources; this distinction is lacking in the

3 The afference is the (sensory) input and the efference the (motor) output; outgoing efferences also leave processable traces.

confused protorepresentation afforded by the child's internal object 'breast'. The relevant afferences must also be kept distinct from the concomitant auditory inputs, or there will be a 'speaking nipple', or a nipple that makes certain noises – as is indeed possible considering the internal object 'breast' and the baby's experience. Yet another distinction concerns concomitant painful sensations: if the baby, while sucking, is pricked by a pin or has a stomach-ache, the afferences concerned are not separated, since the protorepresentation resulting from the internal object 'breast' does not contain such an organization, so that the result is a 'nipple-pricking-the-bottom' or a 'nipple-causing-a-stomach-ache'. Hunger itself, as a painful visceral sensation, is not distinguished from the afferences that identify the breast, so that a hungry baby experiences a devouring bad breast.

5 Other sensory inputs must be not only kept separate, but also at the same time linked to the object identified as the feeding bottle – such as pre-sucking visual inputs. On this basis, the system must perform the figure–ground organization (item 1), recognize elementary geometrical shapes (item 2), recognize the precise object concerned (item 3), and distinguish the relevant afferences from others, which may also be visual or stem from other senses. This entire organization must be kept separate from the tactile-proprioceptive-muscular organization, yet combined to identify one and the same object, which is therefore 'known' as unique. A representation of this kind allows the appropriate adult perception of the real bottle. A protorepresentation without such separation, identification, processing and linking of the various memory traces, while at the same time distinguishing between them, will lead instead to the perception of a 'seen breast' as a different object from the 'sucked breast'. That is possible if this cognitive function is experienced on the basis of protorepresentations resulting from the internal object 'breast'. The same applies, with regard to the formation of a representation used for perception, to afferences and motor efferences relating to the grasping of the bottle – otherwise, if the system works with the protorepresentation resulting from the internal object 'breast', there will be a 'grasped breast' that is not the same object as the 'seen breast' or the 'breast-in-the-mouth'.

6 Afferences must, in addition, be constantly 'calibrated' against motor efferences, to permit the essential distinction between what an object causes in us, by signalling its presence to our senses and making itself known, and what we do, on the motor level, to the object, thereby modifying the quality of the afference that makes it known to us. For example, the recognition of a teat must be calibrated against the quality and quantity of the motricity exerted on it; otherwise, the intensity with which we grasp a given object will cause us to perceive it as varying in its hardness. Thus, the recognition that a liquid is being swallowed depends on the

afferences involved in deglutition being distinguished from the efferences of the motricity thereby entailed; in other words, the act of sucking must be distinguished from what is being sucked.

7 To construct a representation appropriate for the realistic perception of objects, the mind system undertakes a further processing operation, on the traces of the various afferences it has retained. This involves the placing of the object outside the subject's body, so that the type of relationship between body and object can be established and subject-to-object actions can be distinguished from object-to-subject actions. If this does not occur, the external object will not be distinguished from the subject's bodily self, and actions performed will not be distinguished from actions undergone. This may indeed occur if the mind works with the internal object 'breast'; object and subject, the inside of the body and the outside world, or passivity and activity, may then be confused.

8 Finally, all the information from the various senses is kept separate, in its traces organized as described above, from information stemming from the inside of the system – that is, from all the incipient mental products whose traces, like those from external afferences, constitute information that must be appropriately processed. In this way, if the incipient mental system is producing an image of some kind, this must be kept separate from external perceptual traces. It may be imagined that, if a baby is fantasizing a teddy bear, his 'breast' will then, in the absence of certain organizations, also be a teddy bear. More simply, if a neonate is recalling auditory traces – the auditory structures, after all, come into being at a very early stage of fetal development – his 'breast' may also include the recalling of these traces. In more classical terms, if the incipient mind system is able to produce a state describable, by analogy with adult states, as 'anger', 'hate', 'aversion' or the like, the information on this internal state will probably be processed not separately from the sensory traces, but in integration with them, as a 'hating breast', an 'angry breast' or, ultimately, a 'hostile breast'. Everything we call 'primary affects' may be processed *together with external afferences*, giving rise to the strangest phantasms, in turn endowed with these affects and these incipient 'intentions'. This process will be described in greater detail later, in our examination of the 'bad breast' and of feelings of persecution – as well as of the 'loved object', here recalled as a possible 'loving object'.

By proceeding backward from the analysis of the organizational stages of the adult perception of a breast, we can reconstruct the probable steps in that organization in an infant, and identify the successive operational capacities acquired by the mind system, starting with the first protomental operations. The following sections and chapters will be devoted to this description.

4.6 Does pain underlie the beginnings of the mind?

The dilemma of the beginnings has already been mentioned (Section 1.6): if perception is the attribution of meaning, the prior presence of a mnemic signifier is necessary for the perception of, or the attribution of some kind of meaning to, a set of incoming data. We are thus confronted with the problem of how the first signifiers – that is, meaningful sets of traces of inputs, whether external or of bodily origin – are formed. In a hypothetical zero state, data emanating from inside the mind system ought not to be considered, since this system does not as yet exist. The only way out of this impasse is, notwithstanding our experiential approach, to resort to the hypothesis of mechanisms of biological origin that confer the first meanings on certain inputs. These mechanisms would act as primary signifiers. This hypothesis is borne out by certain examples in the field of reflex activity. If stimulation of the buccal mucosa gives rise to a sucking reflex, it may be postulated that this reflex circuit, even if peripheral, may somehow be recorded at central level; this implies that a given input from the tactile receptors of the lips has had some kind of rudimentary meaning assigned to it, so that the information must have been read by a biological protosignifier. The same applies to the other reflexes present at birth – and also to those observable in the fetus. Painful stimuli and nociceptive reflexes are particular examples. Unlike pleasurable experiences, whose activation calls for acquired functional capacities, an unpleasurable experience, whose prototype is pain, is assigned meaning biologically, by the biological apparatuses of specific receptors and spinal neural mechanisms. A fetus, at a very early stage of development, already seems to attribute meaning to certain stimuli, which are thus experienced as painful.

In our search for the first signifiers, we are therefore bound to turn first of all to those which, although activated by experience, are constituted by biological mechanisms. Here we may seek the prime mover which triggers the capacity for assigning meaning – that is, which opens the way to the acquisition of operational capacities. The complexity of the first operations that may be 'attempted' in the search for better operational solutions – that is, operations effective for the purposes of learning – is manifest: how many 'erroneous operations' can be attempted by the incipient mind system before a degree of adaptation[4] is attained? Here again, the fundamental consideration will be the essential survival mechanisms used by human beings to confront hunger, thirst, physical pain, cold, the need for oxygen, and so on. For these purposes, the innate mechanisms perform the first functions of assigning meaning to certain stimuli that would, if not confronted in this way, cause the

4 'Adaptation' is here used in the sense of a solution appropriate to reality (Hartmann 1939).

organism to die. The common denominator of this first meaning will then be the degree of threat presented by a danger.

Great difficulty attaches to these possible reconstructions of the origins of mental life, particularly if they are held to date from birth rather than the months of gestation. In this study, on the basis of the best available information and the tradition of psychoanalytic research, I shall trace the origins of the protomental to neonatal experiences – those connected with the 'breast', or rather those from which the internal object 'breast' originates. The account given would of course differ in many respects if a different starting point were adopted – specifically, that of fetal life.

A few hours after birth, a state of discomfort arises in a baby because the level of nutrients contained in the umbilically transmitted blood has fallen below a certain limit. This is something uneliminable and 'bad' that suddenly appears at the 'zero' point of his mind, of which he can by no means be said to be conscious, even though, at biological level, this situation must be vital to the onset of certain registrations that will subsequently give rise to 'functions'. Neonatal motricity in the event of hunger shows that the biological (biochemical and visceral) signal has some kind of 'mental' effect. In general, when a baby is hungry, the nipple or bottle is placed close to his mouth, triggering an innate neurological mechanism whereby pressure on the buccal mucosa gives rise to a series of muscular contractions in the mouth and throat and liquid is ingested (through the sucking reflex, which is already present in the uterus). The result is a change in the state of discomfort and the disappearance of what may have been 'felt' as bad.

An initial connection between an experience of pain and that of sucking may reasonably be supposed to arise in the neonatal mind: the receptors simultaneously emit tactile signals from the mouth and tongue; gustatory, olfactory and thermal signals; and sets of visceral afferences from the stomach and from the chemoreceptors that record the haematic variations in the transition from the fasting state to that of a sufficient supply of nutrients. The afferences received during and immediately after sucking can begin to assume an initial meaning only if compared with another 'set' of afferences – namely, those biologically registered as 'pain' due to the fasting state. In this way, the two 'sets', when contrasted with each other, can take on a first meaning, which is protomental.

Such a connection, and hence the assumption of an initial meaning, must involve the possibility of memorization. The term 'memory' can legitimately be used when a first set of afferences acquires a first meaning of its own; does this constitute the origin of a first 'internal object'? The sucking-related afferences certainly acquire meaning in so far as they are bound up with the unpleasurable experience of hunger and with the fact that this experiential state can disappear, precisely, upon sucking. So the first 'internal object' of the mind seems to be one associated with unpleasure and the possibility of its elimination; it is therefore a 'bad' object, which must be expelled.[5]

This object is made up of afferences of which very few belong to a real external object; most of them relate to experiences of bodily functions, confused both with each other and with those pertaining to the external object, without distinction between inside and outside. At this level, then, there is no question of a mental operation performed with a view to securing, or approaching closer to, a real object that will get rid of the discomfort. The only operation possible will be somehow to eliminate the painful internal state. It may be postulated that the neonate will use the first organized traces of afferences precisely to eliminate the pain-signifying organization. He will use the first memory traces to rid himself of the first 'mental' aggregates forming from these. The first set to assume a mental guise – the breast – can thus be seen as an object to be eliminated because it is painful; and, for its elimination, the mind can use only the more elementary sets of afferences with which it formed that object. In other words, the mind will attempt to dismantle it so as to return to meaningless afferences.

In the above conjecture, the first meaning proves to be based on a primary contradiction: the signals of discomfort are coupled to the external afferences concomitant with the disappearance of the discomfort itself, thereby forming the first signifier. It follows from this ambivalence that this first 'internal object' tends to be abolished when the biological signals of discomfort reach a certain intensity. This gives rise to the first, perhaps fundamental, dilemma within the nascent information-processing system: which operation is right and which is wrong? Should the first engram, which contains a painful element, be formed, retained, memorized and used, or should it be erased in the very process of its arising? In my view, this dilemma can be associated with the post-Bion psychoanalytic notion that the toleration of mental pain is essential to the attainment of thought. Yet the outline given here surely concerns far more rudimentary levels, even if there are similarities with later stages.

Hence, the first object to assume a clear mental meaning – the first internal object – does not correspond to any definite real object, or to any clearly circumscribable and describable internal experience; however (or perhaps precisely for this reason), it is of fundamental significance in that it is the first object to appear in a system at zero level – that is, an objectless system. Moreover, since this first object is painful, its appearance can be likened to

5 Little information is available on a possible precursor of this bad object, which probably arises out of the experience of hypoxia after the cutting of the umbilical cord and the onset of respiration. However, respiratory fantasies and consequent respiratory symbolization may be invoked (Imbasciati 1983b). Like symbolizations of (auditory?) fetal experiences, this may well predetermine the modes of oral symbolization.

the emergence of a bad entity from the nothingness that knew no pain, as if an actual, terrifying 'phantasm' were not only the mind's first object, but also the first mental 'subject', or as if an ataractic nothingness were to give way to a first existence of suffering and fear. The first function of the nascent mind will be to eliminate this first bad entity that has arisen within it. The first function of the mind may, paradoxically, be said to be self-destructive.

In effect, the neonate – or perhaps, better, the fetus – which up to a certain point has always lived in a state of beatific nothingness, is suddenly assailed by a bad phantasm looming on the horizon of its nascent mind. Since this emerging bad entity is undefined except as regards its painfulness, since it appears gigantic compared with the nothingness of the mind at zero level, and since, finally, it is invasive in character, given that hunger is uneliminable and progressive, it can be conceptualized – in adult terms – as a terrifying, invading evil spirit, formless and all-pervasive. The language used to describe this phenomenon is necessarily figurative, so that it can convey only a small part of the neonatal experience – which is in fact ineffable, because it precedes any category not only of language but also of logic itself. However, we shall for convenience use precisely these terms – phantasms and evil spirits – to describe what probably takes place in the baby's mind in terms closest to our adult understanding. We shall therefore refer metaphorically to the need to 'exorcize' this primordial bad demon.

Chapter 5

Levels of protomental operations

5.1 Exorcism and autotomy

The hypothesis of primal, biologically determined, danger-signalling signifieds has enabled us to describe a first level of protomental operations – namely, the formation of a primitive bad internal object. This first protomental entity is thus an elementary trace providing a basis for a first reading of incoming data, and hence the first protomental signifier for assigning a first meaning to experience. Starting from this primitive internal object, let us now attempt to reconstruct further levels of protomental operations.

For heuristic purposes, eight successive levels of basic protomental operations can be identified. The *first level* was described in the previous chapter, and comprises protosignifieds linked to biologically predisposed signifiers and the emergence of the first bad 'object'. The *second level* of functioning will therefore be that described above, with some poetic licence, as exorcism of the bad phantasm. The term 'exorcism' is used here as a metaphor to denote the 'magicking away' of something bad, just as the Devil is supposed to be exorcized from the body of the person he has entered and possessed (the word is derived from the Greek *exorkizein*, meaning 'to cast out'). It is an effective description of the formless, ineffable entity that probably takes shape in the neonatal mind. To describe this process, it must be postulated that, on the basis of repeated experiences, a comparison takes place between subsets of traces of afferences that have entered into the composition of the first signifier, or engram – namely, those concerning signals of discomfort on the one hand and, on the other, the concomitant afferences of sucking (accompanied by disappearance of the discomfort, although this may well not be relevant). According to this hypothesis, some of the first-level memory traces are compared with others from the same set, giving rise to a different assembly of the traces that convey the meaning 'pain'. In other words, for this operation, the same memory traces must be used as those belonging to the act of sucking and to the disappearance of hunger in order to combat hunger. This constitutes a hallucinatory use of a memory trace to exorcize the looming evil spirit (see Section 4.6) of hunger. It is hallucinatory because a kind of image of the feeding breast is used as if it were an actual feeding breast coming to the aid of the hungry child; expressed in terms of a later developmental stage, an object that could have been 'remembered', or, if you will, 'imagined', is used in hallucinated form – that is, it presents itself as if it were being perceived. In the absence of any distinction between inside and outside, the three categories can coincide in a single, confused, hallucinatory type of 'presence'. In effect, the neonatal mind, 'quasi-remembering' something to do with the disappearance of hunger, believes that it is creating a feeding breast.

Yet this is an adultistic conception of what might happen in the mind of a newborn baby. Moreover, from a psychophysiological point of view, it remains unclear why a set of traces assembled precisely in order to constitute a first signifier should be reconfigured in such a way that two parts of these traces are separated out so that one is opposed to, instead of being fused with, the other, thereby forming another signifier through the hallucinatory use. On the level of these highly primitive operations, other, even more primordial, signifiers are probably involved – namely, engrams already formed in the fetus. The fetal mind certainly possesses tactile proprioceptive-auditory-vestibular-motor engrams, which do not necessarily possess the character of unpleasurableness – that is, of the 'bad'. A possible hypothesis, in fact, is that precursors of 'good objects' have formed in the fetus, and constitute the basis for differentiation of the trace of the 'first bad phantasm'. The fetal organism might already have learned to distinguish good from bad.[1] The quality of these first differentiations of engrams may perhaps even modulate subsequent development, up to the onset of the depressive position.

This 'hallucinatory' use of the memory trace certainly involves the same motricity as that employed for exorcism: a hungry neonate attempts sucking movements even in the absence of the nipple or bottle. Alternatively, he may suck his finger. Like the former, simpler action of even minimal sucking movements, this seems to have the hallucinatory significance of evoking or creating a hallucinated breast to exorcize the bad phantasm of hunger. The trace of the motor program here deployed includes a meaning that is used to modify that conveyed by the afferences of hunger. In response to the 'bad' presence of hunger, a first mental operation is established, whereby moving the mouth, swallowing and sucking serve to cast out the malign presence. However, this operation does not involve satisfaction or an experience of need, let alone a recognition of need. Nor can it include any element of recognition of an object that might satisfy the need. There is merely a negative experience bound up with the necessity for something bad to disappear, even if this ultimately entails the disappearance of a first incipient function of the mind.

1 As stated earlier, the fetus is found to have a clear 'perception' of anything that constitutes a danger to it.

The evil spirit to be cast out might in fact be an embryonic perception of a state of need on the part of the baby himself – namely, the first warning signal that something indispensable to his survival must be supplied. The hallucinatory use described above entails the suppression of this vital signal, and of a survival-directed function. The more successful the hallucinatory use, the more this function is suppressed – that is, a part of the mind, and moreover precisely the part that began to function first, is abolished, or, in effect, split off. An incipient function, which could have been used to seek something and which is developed for the perception of a satisfaction-providing object, is destroyed because it emits a signal of suffering that cannot be received as such and must therefore be neutralized. To avoid pain, the mind self-destructively anaesthetizes itself. Many authors – in particular, Meltzer and Bion – have described clinical situations in which the mind, where wholly or relatively incapable of tolerating psychic pain, performs destructive, or anticognitive, operations, and have attributed this action to primitive processes. We agree: at the earliest stages of the mental apparatus, there is no possibility of tolerating an unpleasurable signal, so that a mind, at its initial level, will function precisely by not tolerating any unpleasure and hence also the functions that signal it; paradoxically, it will function by self-amputation of its incipient developing functions. This is the principle we call *autotomy* (from the Greek *temnein*, 'to cut').

Because the mind initially functions in accordance with this principle, the onset and development of mental functions is the complex outcome of a struggle between the growth of vital functions and the immediate, virtually simultaneous tendency to destroy them. This situation can be seen as exemplifying Freud's nirvana principle (Freud 1920g, 1923b, 1924c), as embodied in his theory of the death drive.

5.2 The 'good' *and* the 'bad'

Let us now consider a *third level* of protomental processes, which results from the development of the first two in accordance with the vicissitudes of the reality encountered by the neonate.

Whenever the baby's mind experiences the evil spirit in consequence of hunger, it will set in train the action of exorcism, by a process that may assume the protosignificance of sucking – imagining, so to speak,[2] that he is sucking. While the baby is performing this operation, the actual breast or bottle may really arrive, provided by the mother, or else it may not arrive. In

2 Here and elsewhere, recourse must be had to circumlocutions in order to describe in words phenomena that are actually ineffable; this point should be borne in mind so as to avoid a spurious adultistic conception of protomental phenomena.

the former case, the exorcism will have worked – not because it works in reality, but because, by coincidence, the mother came along at feeding time; but, in the latter, the child is confronted with a potential demonstration of the uselessness of his exorcistic mental operation. If the exorcism does not culminate in the arrival of the mother with food, the baby finds that his attempted mental operation is called into question, and therefore experienced negatively: after all, the hallucinatory anaesthetization cannot persist beyond a certain point, given the increase in hunger-related afferences. Beyond a certain limit, the bad phantasm of hunger can no longer be masked by the hallucinatory use of a protomemory associated with its disappearance. Confronted with the ineluctable progress of the evil spirit's invasion, the child's mind is swamped, and the first set of afferences constituted for memorizing a disappearance of hunger – that is, the exorcistic instrument, in effect perhaps a hallucinated first good breast – is now incorporated into the painful situation of hunger itself. Furthermore, having initially been experienced as an aid in fighting the enemy, this instrument, now that it has failed, is equated with that enemy, with consequent feelings of, as it were, rage, hate and betrayal. The beneficial hallucination, so to speak, changes sign and becomes an integral part of the monstrous spectre of hunger.

The spectre is thereby reinforced, because it is due no longer to hunger alone, but to an unforeseen deviation which, in adult terms, might be called disappointment, chagrin, rage, hate and terror, entailed by the failure of the exorcism. The child's mind, in this first confrontation with reality, is enriched with experiences, or protofeelings, that are essentially painful – that is, bad. A contradiction has presumably arisen in the information-processing system with which the concept of mind and its construction are here equated, thus precluding a solution. In adult terms, the mind is engaged in a futile struggle with the instruments it seemed to have constructed. The situation is, as it were, a precursor of states describable as 'rage', 'chagrin', 'hate' or even a preconception of falsity, the exorcism having proved to be a 'traitor'. Might this be the origin of a primitive syntony of hate, evil and falsity? Therefore, the mental system must seek other solutions. These will be possible if the coming of the real breast alleviates the baby's 'despair', or if some other event read as good (fetal auditory engrams?) intervenes. Conversely, the concomitant occurrence of another kind of painful event (an irritating or nociceptive stimulus or physical trauma; here we are entering the field of pathology) may trigger a tendency to obliterate any pleasurable perception, general anaesthetization or, as defined earlier, autotomy. Having cried his eyes out, the desperate baby will go numb. The resulting, sometimes severe symptomatology, as recorded in infant observation studies, may presage psychotic events (Tustin 1972; Mahler, Pine and Bergman 1975; Meltzer, Bremner, Hoxter, Weddell and Wittenberg 1975; Lebovici and Kestemberg 1978) or future mental retardation.

In this struggle, the act of sucking itself is equated with the bad phantasm and is fearfully avoided: a child desperate with hunger may refuse to suck, or alternate between refusal and angry sucking. The origin of severe cases of food refusal may lie in such situations.

If exorcism fails, the child may face disaster, with a progressively more terrifying bad object – a bad breast – that crushes any 'mental' attempt. This bad breast is the primal signifier of death, not only owing to its biological connection with hunger, but also because it kills off the first, nascent psychic life. Conversely, if the mother arrives in time, the mind will reinforce the experience of its own capacity to exorcize the evil spirit, with the accompanying experience of being able to create – albeit, as we know, in hallucination only – something good to set against the bad. In other words, a rudimentary good breast is formed. This, of course, is in no way attributable to any reality or to an external object (a neonate certainly has no recognition of the fact that the mother feeds him), but is experienced as autochthonous, belonging to the self. It is in fact the initial embryo of a self – that is, the first 'set' of afferences that not only is 'mentalized' but remains in the mind and, being retained there, constitutes its primary nucleus.

Paradoxically, then, while a child's first truth is bad and perhaps disastrous, his first 'good' may be said to be hallucinated and false. This clearly indicates the fragility of the incipient mind: it cannot endure the truth, and must administer a deception to itself. However, the possibility of hallucinating also arises on this third level, and the hallucinated mnemic entity may come to form the first good internal object. The *capacity* for hallucination, on the other hand, arises precisely out of an initial interaction with reality. What is external to the mind – the mother's intervention, or the real external breast – is erroneously experienced as something of the baby's own. So if the mother arrives too promptly, the baby may thereby be led not to recognize reality but instead to fail to recognize it. Sucking will then never become an act directed toward relating to the outside world, nor will the mouth become an organ of knowledge, as it does at a later stage in normal children; the baby will, as it were, be unable to 'learn' to suck, but will continue to use the motor reflex of sucking, endowed with merely hallucinatory, rather than operational and relational, mental meanings.

This suggests the existence of an optimal time lag between the onset of hunger and the provision of food, or of the breast – 'optimal' in terms of the learning and knowledge of reality. If the baby has to wait for a while, but not too long, between the commencement of his state of hunger and the mother's intervention, he will be able to observe that exorcism does not work, but will not feel overwhelmed and annihilated by an over-large and crushing bad enemy. Seeing that exorcism does not work and that a different mental operation must be attempted, he will at the same time be sustained by the inner presence of the modicum of 'good' contained in his hallucinating, enabling him to seek a greater, and truer, good in external reality. Otherwise, he will be destroyed by badness, and will be unable to recognize anything good or any help: a hungry, angry baby refuses the breast, sometimes with the serious consequence of refusing any form of food, or vomiting it up; later, he may refuse all contact with external reality.

If the intervention is too prompt – for example, if a mother anticipates her child's hunger – the child may experience the state of hunger either insufficiently or not at all, and thus remain attached to his prototype of the good breast, which, however, will have been experienced only in hallucination. This may impede his future development toward the knowledge of external reality and toward learning. He might then become fixated on omnipotent experience, in which everything is felt to have been created autochthonously by his own self, and a reality distinct from and independent of himself is difficult to recognize. In this case, the traces of afferences relating to the act of sucking remain invested with hallucinatory meanings, instead of assuming the more appropriate meaning of an action directed toward the discovery and knowledge of the outside world, on the basis of an internal need to be satisfied. Here again, even the experience of the baby's own need may be compromised: in other words, the afferences from the state of hunger, not having been appropriately experienced, may tend to remain meaningless, so that he cannot readily transform them into something 'mental', experienced as a need of his own. The child does not feel hunger, and this fact may presage future mental difficulties in recognizing his own needs and thus transforming them into wishes. Again, hunger-related afferences may in these circumstances tend to remain attached to an experience alien to the self – that is, to be constituted as an alien spirit, which the mind believes it can readily cast out, given the efficacy of its own hallucinatory exorcism. In other words, this situation favours the persistence of primitive experiences of hallucination and alienness in relation to something that ought to be recognized as an internal need; and if this need is experienced as alien, it will forfeit its own specific significance as a need, so that the subject will have difficulty in recognizing parts of himself.

Therefore, if an excessively long time elapses between the onset of hunger and feeding, the baby is likely to be overwhelmed by afferences too intense to be organized as sets signifying something that is happening; in contrast, if the interval is too short, the mind may well be insufficiently stimulated to recognize both external and internal reality, and the afferences will remain organized in signifieds inconsistent with reality. The optimum length of time required to avoid both of these risks cannot, of course, be readily quantified. In other words, it cannot be asserted that it is good for a child to remain hungry for 2 or 3 minutes rather than for half an hour, or that it is better to wait for him to cry before feeding him, instead of being guided by other signs of his hunger. The optimum interval varies widely from one neonate to another, and also depends on the relationship established with the mother and on the communication (see Section 10.4, on the function of reverie) that takes place during the interval, especially in the pre-feeding period.

Hence, the product of the third level of protomental functioning may be described as a capacity which, while hallucinatory in nature, indicates the stabilization of a memory trace, experienced as the first good object – perhaps definable as a precursor of the self.

5.3 'Inside = good = self' and 'outside = bad = not self'

Let us now analyse a *fourth level* of primitive mental operations. The average neonatal experience includes a kind of offsetting of situations of despair, making for disintegration of the mind, against others that reinforce and promote mental operations. This can be conceptualized as a dialectic between the first mental images – internal objects – which are conventionally described as the good and the bad breast. These two objects alternate continuously and the child must learn to distinguish them. This is no easy task, because they do not correspond to two distinct objects in reality, or to two experiences that can be clearly separated from each other, but in fact stem from the same traces of buccal afferences and of the unpleasurable state of hunger, as well as of other afferences. The mental system must learn to distinguish good from bad on each individual occasion, according to the situation of the moment. Later, when it has acquired the capacity to distinguish between external and internal mental experience – that is, when an external reality of objects distinguishable from their mental representations begins to take shape – it will have to learn to determine whether those external objects are good or bad, given that they may coincide with either unpleasurable or good experiences (and hence be experienced correspondingly). In other words, within one and the same set of afferences, it must first distinguish between what is good and what is bad, and subsequently learn what is inside and what is outside.

As we have seen, the bad breast assumes the form of a malign presence, whereas the good breast is a hallucinatory presence used to banish the bad presence. That is to say, whenever the child becomes aware of an unpleasurable experience, he will activate sets formed by earlier traces in order to eliminate these traces or keep them at bay. There being no distinction between memory, image and perception, or between the external reality of things and the thought of them, and indeed no conception of how thought can act on reality, the child may be said to be trying to use his proto-memories as instruments to eliminate unpleasure. The good breast, when it begins to exist, is felt to be something of the child's own that is always at his disposal. Conversely, the bad breast, as a proto-memory of the experience of non-satisfaction, is something that must be eliminated, and is therefore immediately felt to belong to something different from self, so that it is attributed to something other, external and alien. This, then, is the foundation of the primitive distinction between a primitive self and a non-self.

We know with our adult minds that the memory of hunger stilled is something of our own, like the memory of the experience of hunger and the actions deployed to relieve it. For a very young child, however, the equivalent of the memory of unsatisfied hunger may be an evil spirit alien to himself and not belonging to his mind; the unpleasure is not a state of non-satisfaction of his own, but takes the form of something different-from-self, or alien. This process – experiencing something that is actually produced by the subject's own mind as not belonging to the self – is called primitive projection or expulsion. In this way, the bad – indeed, anything bad – takes the form of something alien, of non-self.

The opposite process may also be observed. Here something actually originating from an external object – afferences of satiety following the coming of a real external breast – is organized as if it belonged to and stemmed from the subject himself. This process could be described as primitive introjection. The mind's first mode of distinguishing between good and bad now comes into being: what is good is experienced as self, while what is bad is organized as non-self. At the same time, a first distinction is drawn between inside and outside: the good = self is inside, while the bad = non-self is outside. Of course, like its predecessor, this distinction is erroneous – that is to say, it does not correspond to factual reality. Nevertheless, it constitutes a mental operation of paramount importance for a neonate, laying the foundations for the entire subsequent organization of other mental operations.

The distinction can be summarized in adult terms as follows: 'Everything that is good is inside me – indeed, is myself – so that everything that is in me and that I am is good; on the other hand, everything that is unpleasurable does not belong to me and is outside me; let us hope that it does not exist, but, if it does make its presence felt, I shall do my best to cast it out into the void again.' Certain unconscious pathological experiences of adults can be discerned in this conception. The distinction between inside = myself and outside = other will come to be extremely powerful, so that even when the child has performed other, more complex operations, he will always have this tendency to attribute all good things to himself and not to attribute anything bad to himself. The child will always be led to believe that all good experiences belong to himself, while bad ones are alien and external, so that development will take the form of a dialectical process in which the first incorrect mental operations alternate with later corrective ones.

This sense that good things belong to the self, as 'me' and later something of 'mine', suggests an explanation of why the mind constructs itself progressively, why there is a tendency for thought to develop and why excessive frustration undermines the very foundations of the capacity to think. The first mental operations, to the extent that they serve to cast out the 'bad' – that is, in reality, to endure the pain of waiting – are felt to be good and to belong to the self, so that the individual will tend to keep them for himself and to develop them more and more: mental operations, including those which, while inappropriate, afford momentary relief, are the first 'good' that a child possesses. They are themselves a 'good object'. Probably for this reason, he will tend to preserve them inside himself – that is, to preserve something mental that has successfully organized his own suffering. It will then be more readily possible to transform the afferences into mental organizations endowed with meaning – even if they do not match reality – so that they can be used to confront, if not reality itself, at least its distorted mental appearance. That is a possible reason for the retention and preservation of 'mental operations' by the mind, in spite of the primitive tendency to erase and destroy them as mentioned above, and for their potential to undergo further development. This situation may be one of the roots of the first mnemic processes, and its intensity may underlie the subject's greater or lesser capacity to remember. Later, perhaps, it may form the basis of trust in the subject's own capacity for thought, as well as of its efficiency.

That may be why mental operations in themselves come to be experienced as 'good', so that knowledge, which is mediated by mental processes, intrinsically appears as good. This is so even if the first knowledge is inappropriate and false, designed more to alleviate unpleasure than to know the truth.[3]

5.4 Mental operations and reality testing

A neonate can verify the immediate utility of the first thought schemata for the purposes of his momentary mental economy, but cannot test their validity with respect to external reality, because they are used essentially to exorcize the 'bad'. Testing of the validity of the mental operations performed – of their suitability for representing actual external reality, for the purpose of knowing it so as to act on it effectively – can begin only when the exorcism of the enemy is proved ineffective. However, this is not easy, because exorcism is deployed not only in the presence of a bad experience capable of invalidating it, but also – and perhaps principally – when the intensity of the experience is so weak that it is masked by the hallucinatory manoeuvre, and probably, too, when the child is not hungry: the observation of newborn babies indicates hallucination of the breast independent of feeding rhythms. In all these cases, the neonate experiences a confirmation of the efficacy of his exorcistic operation, but it is a false confirmation, not invalidated by the intervention of a reality that imposes itself as different. Exorcism thus appears to function in the neonate's mind even if food is offered to him promptly, or perhaps too promptly, so that he in effect believes that he has created it himself and cast out the enemy. In all these cases, the child is, as it were, made to have the experience that there is always something to hand,

3 If such forms of knowledge persist in an adult (as in narcissistic or paranoid states), he will cling to them with particular stubbornness notwithstanding contrary evidence from reality.

created by his own thought – indeed, generated within it – that causes his discomfort to disappear.

Conversely, when the baby is forced by persistent hunger to experience what, for an adult, is the absence of something (that is, when he is faced with a situation in which he might be led to recognize what he needs in an external entity that is absent, and to identify his own state as need), and to experience the non-immediate availability of something that therefore proves to be independent of himself – that is to say, in all situations where he might be led to feel that the exorcistic operations have failed – there are two contrary possibilities. On the one hand, the system is induced to search for other mental operations, but, on the other hand, the encounter with a reality different from that hallucinated might be incommensurate with the mind's capacity at the time, so that what would have been an object to be recognized as different-from-self and to be searched for is instead experienced simply as a non-self to be rejected. A state comparable to rage and fear then leads the mind to fail to recognize the reality that was about to be revealed, and a fresh attempt at exorcism stifles the invalidation of the exorcism that was on the point of being experienced. The more the child's mind is traumatized by the unpleasurableness of hunger-related or other afferences, the more frequently this will occur; and the same applies, later, to abandonment and to other unpleasurable states such as concomitant painful physical stimuli.

The fact that there is not always something at hand to cause its own states of unpleasure to disappear might be the starting point for the mind's eventual construction of an experience that matches external reality sufficiently closely – namely, the experience of the existence of objects that may be momentarily absent. While we adults know this, for a child the path leading to the full experience of this fact is exceedingly arduous: this *something-not-readily-to-hand* is experienced as hostile and bad, and the absence of a beneficial object is read as the presence of a maleficent object. All clinical infant studies see the capacity to perceive an absent object as of later onset than the perception of presences. The absent object is a present 'other', 'bad' object. The element of experience whereby successful exorcism ought to have been distinguished from failed exorcism is felt to be the bearer of the 'bad'; in adult language, it unleashes rage. An adult knows that this element is bound up with the presence or absence of a real object: what distinguishes successful from failed exorcism is the presentation or otherwise of milk. An infant, however, does not know this; just as he is stimulated to recognize the existence of a reality outside himself and beyond his control, so he is frightened by and enraged at this variation in his experience, which he attempts to reject and disavow, by placing it outside himself and accommodating it not in the existence of a real object but, as before, in an imaginary enemy to be cast out.

The first experiences that may lead a child to the knowledge and understanding of the existence of a real object different from himself, and also different from his mental representations, are contradictory: on the one hand, they include the possibility of distinguishing outside from inside, and hence hold out the prospect of actively controlling what is outside; while, on the other hand, they involve frustration, perhaps taking the form of rage at not being able to do so immediately. External reality therefore presents itself as endowed with unpleasurable qualities to be rejected and, until the discomfort can be endured, the child will continue to repudiate that reality instead of recognizing it: until the mind is strong enough, it is more economical to persist with exorcistic 'thinking'.

When hunger is very intense and exorcism does not work,[4] the individual is compelled to seek another route, and hence to tolerate the bewilderment of not being able to control at will the variation of experience that tends to make for the real representation of the object. However, it takes a long time for this route to assume the form of the realization that the object cannot be made to appear at will, although it can be remembered and imagined. Similarly, a prolonged and laborious state of contradiction – the possible prototype of the affect we call rage – will be inevitable: successful exorcism leads to the belief that everything can be obtained magically, as if produced by omnipotent thought, while its invalidation by reality proves unexpected and cruel. When a young child recalls things that might banish the bad presence but finds that his exorcism does not work, he might be able to discern the difference, whatever the variation of experience concerned. However, this variation, which is bound up with the presence of the real object and hence beyond the child's control, is unpleasurable, so that he is led to reject it, and to go on living as if the whole world existed inside himself: everything is thought (or rather pseudo-thought), imagination and delusion, and can therefore be produced utterly at his whim.

Whereas the first mental operations may be felt as 'good' (see Section 5.3), it is precisely those whereby reality might be recognized that may be experienced as 'bad'. Knowledge then presents another face – that of persecution – and thought takes the form of a possible epistemophilia, intrinsic to the construction of the mind and ambivalently directed toward both illusion and reality. The tendency toward reality entails anguish and suffering. The child is led to reject external reality, although a powerful need urges him to recognize it; the new and external is seen as an enemy – and, moreover, not only is the enemy alien, but also the other-than-self is an enemy. The trend toward the recognition of external reality in the form in which it presents itself cannot become dominant until this difficulty has been overcome – that is, until the mind has enriched and fortified itself

4 As stated earlier, this description is based on a neonate in normal, optimal conditions. If the situation differs, however slightly, from this norm, as, for example, in the presence of physical pain, the entire course of events described here – in particular, the exorcistic operations performed to ward off unpleasurable, painful stimuli – will be changed.

sufficiently to tolerate contemplating the enemy – in order to realize that it is *not* the enemy.

The mental work required to attain this new realization is enormous, and can be seen as the *fifth level*, or phase, of protomental activity. It is made up of an ongoing dialectical alternation between the first (hallucinatory, exorcistic and autotomic) mental operations, on the one hand, and, on the other hand, a second type appropriate to knowing and acting on reality. Because this phase is so important to the foundation of the mind, it will be described in more detail below, on the basis of the Kleinian conception of the paranoid-schizoid and depressive positions.

5.5 The unpleasure principle: autotomy or cognitive development?

In studying the origins of the first mental operations, we may enquire as to the reason for their constitution. Having discarded the hypothesis of an endogenous energy (the 'force' identified in instinct and the drives), we have sought to answer this question in terms of the problem of survival and adaptation: the mind can then be said to develop so that the individual's needs can be satisfied. The mind can be regarded as an instrument that constructs itself in relation to the survival of the individual and the species, performing a function analogous to, but much more developed than, that of instinct in lower animals, and allowing life to be lived even in more difficult environmental conditions. As a learning system, it has learned to learn progressively more, and in this context also how to transmit the entire system from one individual to another. This is the foundation of its constant transgenerational growth, which has made for the evolution typical of the human species and for the construction of civilization. Human biological life, at least for the species, is not conceivable without mental life, although it may possibly be for an individual in specific circumstances. Man has survived, and continues to survive, by developing his mind. The deep experiences of self-knowledge, and indeed all knowledge, bear the traces of the most crucial problems faced by the species: epistemophilic anxieties have moulded man's destiny (Imbasciati 1979c).

So if the mind arises in order to satisfy needs, what is it that forms primarily, as a signal that they can be satisfied? In this connection, the satisfaction of need has been conceptualized as 'pleasure', and pleasure due to the satisfaction of need has been postulated as the underlying principle of mental organization (Bernfeld 1925). The Freudian tradition has constructed its entire theory of development along these lines, deeming the pleasure principle to be the prime mover of psychic life. In the light of subsequent findings by the psychological sciences and by psychoanalysis itself, this assumption has proved to be aprioristic. It is based on an adult-type vision and a naive transposition of experiences observable in older children to the earliest stages of mental life, as well as on the presupposition that subjective events of this kind can be incorporated in an objective order of causality in relation to the genesis of functional structures. Again, Freud saw in this notion one of the many applications of the neurological principle of energy discharge, on the basis of which he constructed his entire drive theory, so that what is now called into question appeared as an obvious postulate. Pleasure is in fact not an *explanans* of psychic life, but an *explanandum* (see also Chapter 2). It is already a complex psychic event, which cannot be invoked as the prime mover of psychic life.

Examination of the beginnings of mental functions does not reveal any event equatable with pleasure that might constitute an elementary 'signal' for validation of the first operations. It in fact seems more appropriate to conceptualize this discriminating signal as pain, suffering or 'unpleasure'. Fairbairn (1952) and Guntrip (1961) invoked the unpleasure principle long ago. More recently, the Bion school (Meltzer 1978; Bion, F. 1984) has considered 'mental pain' (see Chapter 6 below) as the primordial state of the mind that acts as a signal for development, whether constructive (the onset of the depressive position, mental anabolism or symbolopoiesis) or involutional (paranoid-schizoid processes, 'evacuation' – cf. our concept of 'autotomy' – or catabolism) if the nascent apparatus is unable to endure it so as to process it constructively; that is, if the degree of 'pain' is disproportionate to the capacity of the apparatus. Mental pain is identified as the ultimate, irreducible and uneliminable element of the mind, so that the problem of development becomes that of *containing* it (cf. the concepts of the alpha function, alpha elements/beta elements, and the container-contained – Bion 1962, 1963, 1970). To return to the origins of the mind, to the protomental and to the 'reason' why the first operations begin, an unpleasure principle, or rather an unpleasure avoidance principle, can thus be seen as the initially biological and later 'mental' trigger for the commencement of 'protomental operations'.

On the basis, again, of examples in the field of biological needs, hunger may be supposed to give rise to an experience of unpleasure; and eating, in a hungry individual, to pleasure. On closer examination, however, whereas the assumption that hunger constitutes unpleasure proves justified, it is extremely doubtful that eating in itself constitutes a pleasure, at least for the neonatal mind. The evidence of neurophysiological research suggests a virtually automatic experience of painful stimuli; and the biochemical imbalances of hunger presumably also constitute a powerful afference, which, so to speak, 'forces' a mental experience to be formed, thereby setting in train processes, which may be mental or behavioural, directed toward its modification. This modification is accompanied by the disappearance of the unpleasurable mental experience; and whether this is due to the ingestion of food rather than to hallucination depends on whether the subject is fed passively or is capable of seeking and finding food when hungry. However, the latter possibility presupposes an already developed mind, capable of 'knowing' that this particular state of discomfort disappears if food is sought and found; that is, it presupposes an experience that the neonatal mind is yet to have. Nor does the fact of the subject's being fed passively involve recognition of the satisfaction of a need; indeed, as described above, it may favour the omnipotent, hallucinatory experience that food is created by the mind, or rather that suffering (due to the absence of food) can be effectively combated by the mere hallucinatory activity of the mind. It is therefore simplistic to assume that, to cause the discomfort resulting from a need to disappear, the primitive mind attempts to satisfy the need, for this would imply the existence of an already developed mind. Instead, it may be postulated that the primitive mind, when confronted by a state of unpleasure, tries something intended to make it disappear, and that it opts for the simplest relevant action, which is suppression of the capacity to receive the painful afference – that is, to 'perceive the unpleasure'. A first mental operation in response to an afference pressing to be experienced as unpleasurable may be to reject it; in other words, a first mental operation may be undertaken with a view to annihilating what the biological apparatus seeks to present as the first mental element. The first mental operation may be autotomic.

Conversely, the notion that the ingestion of food is a pleasure to a hungry individual presupposes that his mind has already 'learned' that eating causes the discomfort of hunger to disappear. For the mind at its primal levels of functioning, pleasure cannot be regarded as the primary parameter of mental operations. Pleasure is already the result of mental operations of learning and the use of experiences. As an emotion, pleasurableness presupposes complex processing. Pleasure must therefore be regarded as secondary and subsequent to unpleasure, which is in fact the primary entity presented by the biological apparatus as an element capable of becoming mental. Pleasure as a state stemming from the satisfaction of needs is therefore a mental construct, formed by a mind that has already attained a certain level of development. On the primitive level, the first pleasure may consist simply in the elimination of pain by means of an autotomic operation whereby the receiving mechanism of the signal of unpleasure is destroyed. In other words, on this primitive level, it is less a matter of pleasure than of the elimination of unpleasure, and the quickest way to accomplish this is to perform a self-destructive anaesthetic operation resembling the autistic passive dismantling and fragmenting of the nascent mental organization (Meltzer, Bremner, Hoxter, Weddell and Wittenberg 1975).

Like oral pleasure, a pleasure derived from the excretory functions has been taken for granted; in fact, however, the existence of such a pleasure cannot be presupposed in the neonatal mind. There is merely the disappearance of the discomfort due to retention when the mechanism of excretion operates. Other 'pleasures' stemming from the satisfaction of other needs have likewise been adduced, with regard first to elementary needs and later to more complex ones. For instance, the idea of 'motivation' has been used to denote the intensity with which a need leads to a form of behaviour directed toward its satisfaction. Besides the elementary motivations of hunger, thirst, sleep and cold, that of sex has been invoked. Freud thought sex so important that, enshrining it in the concept of 'libido', he elevated the 'pleasure principle' to the fundamental parameter of primary psychic life. But pleasure is actually a function of an already developed mind trained to know that, among the various ways of eliminating unpleasure, there is a form of behaviour that will intelligently satisfy the need. As to sexual pleasure, its nature is still disputed, and can by no means be assumed to lie in a 'naturally' pleasurable afference. There is much evidence to suggest that sexual pleasure is a complex mental construct (Imbasciati 1979b), and some authors see it as the effect of a reassurance against destructive anxieties (Klein and Riviere 1937). Finally, turning from needs associable to some extent with the biological level to more strictly psychological ones ('psychological motivations' or 'secondary needs'), the literature on pleasure as the underlying principle of need-satisfying behaviour is found to be highly ambiguous, when considered in terms of primary psychic operations.

With regard to nascent mental life, then, little store can be set by the pleasure principle, since the concept of pleasure as yet has no meaning for the mind; an 'unpleasure principle' is a more appropriate concept, since the avoidance of pain is the primary parameter in the efficacy and stabilization of a mental operation.

This, however, raises a problem to which we have already in part alluded. The abolition of unpleasure is essentially secured in two ways. The simpler is a self-destructive operation that eliminates the signal by weakening or demolishing the receiving mechanism, or, better, the mechanism for processing the received signal. The other, which is more complex, points to an operation describable as intelligent, directed toward the actual satisfaction of the need. The problem, then, is to explain how the mind arrives at this second solution, abandoning the first, more simplistic approach of annihilating itself in total ataraxy – which, in the end, is tantamount to death.

5.6 Satisfaction and intelligence: the reality principle versus the knowledge principle

Afferences from unsatisfied biological needs give rise to mental experiences comparable to those stemming from any other unpleasurable afference – such as the painful stimulus of the skin being pricked. Let us therefore take this example as our model. There are two basic ways of dealing with a pricking object. The first and more common is to remove the object from the skin or the pricked part of the body from the pointed object; this behaviour is directed toward the realistic elimination of the pain and the damage it signals. However, if this solution is unavailable, if the agent pricking or irritating the skin cannot be removed or we are immobilized, another method is to try not to feel the pain, either by enduring it or by the use of an anaesthetic.[5] In this case, the anaesthetic is our only pleasure – or rather, it is the only way of eliminating the pain.

Another example might be a feeling of sadness and depression: one approach to the alleviation of this painful state might be to meet friends. However, if this is impossible, because one either has no friends or is incapable of relationships, a possible way of not suffering and of trying to avoid feeling sad and lonely is to become numb; if this cannot be achieved by internal processes, recourse may be had to external assistance, as for instance, by getting drunk. Again, if one is lonely, anxious and in need of deep communication and contact, and these needs cannot be met by intelligent behaviour realistically directed toward their satisfaction, one may resort to drugs, which will blot out these experiences and transform them into false perceptions of happiness and self-sufficiency.

The mind – including the adult mind – always has two contrasting ways of confronting adversity and unpleasure, one directed toward modifying the relevant reality and the other toward modifying the receptivity of the painful signal. If we see a lion and are afraid, we may try to defend ourselves – or we may close our eyes and fall into a trance, like a frog faced with a snake. Although these examples are simple and sometimes paradoxical, they indicate one of the ways in which the mind deals with unpleasurable states – and one that is more common than might be thought. The examples of alcohol and drugs show that the apparatus producing the signal of unpleasure has been demolished 'macroscopically' by external agents of a pharmacological nature. However, there are also 'mental drugs', in the form of exclusively internal processes, that perform the same function as drugs – namely, the avoidance of stress, problems and anxieties that would otherwise cause mental suffering. Sometimes the drug may take the form of action:[6] when faced with certain kinds of suffering, a person may resort to action, with a view less to modifying the cause than to forgetting – for, as we know, unpleasure can be 'stifled' by active behaviour directed elsewhere and causing us to forget. On other occasions, the 'drug' is even better concealed, in unconscious processes whereby a person becomes insensitive to discomfort-producing situations. While often difficult to discern, these processes always have an autotomic connotation. Suffering is sometimes said to have a numbing effect, which means that it can give rise to a kind of armour against unpleasure, but it also

5 The toleration of pain lies on a continuum that includes habituation to pain due to endogenous anaesthetization resulting from the production of endorphins. What analogous production might underlie the elimination of mental pain?

6 Cf. the psychoanalytic concept of acting out, in which, to avoid mental pain and the concomitant insight, the analysand resorts to action, whose primal meaning the analyst struggles to establish.

implies that this insensitivity is generalized, resulting in an acquired 'hardness' in the individual that is seen not as a positive characteristic but as a restriction of his inner resources. In other words, this notion refers to a process of anaesthetization and autotomy of the personality. However, suffering, if calmly accepted and borne, is also said to make one more sensitive and receptive to the problems and suffering of others. This means that the response to suffering may be the development not of an apparatus to defend against the signal, but of an apparatus to process it – that is, of a structure (commonly referred to as emotional resources or sensitivity) capable of mental operations that can only be described as intelligent, which may sometimes include development of the creative capacity or the capacity for human relations.

So it is not true that suffering causes numbness: there is a certain way of learning, in which suffering is tolerated, that leads to an increased capacity for understanding. This is so in particular where the suffering is typically mental or, if somatic in origin, assumes an eminently mental form. The elimination of psychic suffering by means of a defensive apparatus intended to combat signals of pain, or rather of anxiety, is characteristic of certain psychotherapies. The main objective of psychoanalysis, on the other hand, is to make the analysand increasingly aware of and sensitive to his own anxieties, and hence his mental suffering, so that he can develop mental systems suitable for either removing the external sources of pain or tolerating them by working them through in a manner useful to the mind itself; psychoanalysis does not alleviate suffering, but helps the analysand to suffer fruitfully – that is, it teaches him how to understand, tolerate and use suffering.

Still in the context of reality-level, need-satisfying behaviour, two different mental approaches can again be discerned, one directed toward immediate satisfaction and the other concerned with preserving the possibility of such satisfaction. If someone in the throes of intense sexual desire jumps on and assaults the first partner he encounters, he may possibly satisfy his need, but his behaviour will certainly not be very intelligent – that is, directed toward securing a stable way of satisfying his needs. A more appropriate approach might be courtship and the establishment of a loving relationship. Analogous examples could be adduced from the field of ecology. In a famine, if a farmer acts in accordance with the principle of immediate pleasure and satisfies his hunger by eating his seed, he will deprive himself of future harvests; but if he tolerates his hunger, he may ensure himself a better crop next year. Similarly, a need for timber might lead to the indiscriminate felling of all forests, with substantial ecological damage. Again, ill-judged industrial development could pollute the entire planet, so that pandering to the wish for immediate well-being might pave the way for future disaster. Hence, the pleasure-giving object cannot simply be grabbed indiscriminately wherever it is found. 'Things' cannot be appropriated simply because they are plentiful in a given location. One has to consider whether instant use increases or decreases the possibility of future use. The above sociological examples illustrate how a primitive mode of thought can take root in society.

Within the sphere of need-satisfying behaviour, the most appropriate satisfaction will not be obtained by obeying the law of avoiding non-satisfaction at all costs; in other words, the basis of one's actions cannot be a principle of immediate pleasure. Instead, what is needed is a knowledge of reality, allowing one to choose the most appropriate and intelligent behaviour for the satisfaction of one's need, even if this sometimes involves non-satisfaction and the toleration of pain. In every form of mental functioning, from the primal to the most advanced, then, two opposing modes of formation and implementation of mental operations can be discerned: the first is validated first and foremost by satisfaction in the here and now, while the second seeks to know reality for the purpose of identifying how and when it might be possible to satisfy needs – and, moreover, to satisfy them in the safest and most permanent way possible. The first approach bears the stamp of the unpleasure-avoidance principle; the second, that of the reality principle. The first may take the form of behaviour which, while acting on the real world, ultimately proves to be 'not very intelligent'. The most elementary form of such an unintelligent, or, better, anti-intelligent, operation is an autotomic process. Hence, the avoidance of unpleasure leads to death unless offset by the toleration of pain and the knowledge of reality.

Inherent in the reality principle is the recognition of reality; that is to say, the mental system must have constructed a sufficient capacity to recognize reality rather than hallucinating or falsifying it. A minimum level of what might be called intelligence, or the capacity for knowledge, is necessary. Reality makes itself felt not as such, but only to the extent that it is known. Therefore, the reality principle should perhaps more appropriately be called the knowledge principle.

Some authors (e.g. Fornari 1975) who have considered Freud's hypothesis of the pleasure principle in this light have readily been able to show that it must be seen as a manifestation not of Eros, or life, as Freud implied, but of the death drive.

5.7 Avoidance of unpleasure versus knowledge

A dialectic whose terms are the unpleasure principle and the reality principle constitutes the basis of the functioning of the mind and depends on two complementary variables – namely, the objective ease or difficulty presented by the environment and the individual's capacity to pursue satisfaction. In other words, we have a living system of interchange and communication between a personal environment, which may be more or less permeable and responsive to the neonate's requirements, and a person-to-be who constructs himself by utilizing feedback from the environment in such a way as to reduce – progressively, in the case of normal development – the oscillation between evacuation and containment of mental pain, in the sense of an increasing capacity to contain it. In this context, however, it is impossible to lay down an aprioristic normative code of 'good' psychic functioning whereby the rule of pursuing adaptation to reality – as opposed to avoiding unpleasure – is postulated as making for survival and psychological growth, for this would be tantamount to introducing a moral 'standpoint' on a level of consciousness that is wholly lacking, and would constitute an improper transposition. We shall therefore consider the form of mental processing within a given individual's specific experiential situation whereby not only the realistic problem solution but also the avoidance of unpleasure (that is, flight) can be used for life and growth; in other words, in a system in which the permeability of the environment to the individual's needs is extremely limited, the solution of unpleasure avoidance may also prove useful. For this reason, it would be a misunderstanding of the foregoing argument to claim that one must learn to suffer at all costs because this leads to a knowledge of the truth; this is especially not the case for the infant mind. Yet flight, or autotomy, while it may be momentarily useful – like anaesthesia during surgery – can contain within itself a tendency toward self-destruction. Where flight from pain becomes so overwhelmingly necessary as to constitute the only feasible solution, it tends to destroy the subject's mental organization – that is, the apparatus for thinking thoughts and for having feelings (Bion 1962). The mind, in performing an autotomic operation, thus reverses its direction of functioning, 'cannibalizing' (Meltzer, Bremner, Hoxter, Weddell and Wittenberg 1975) the structures already formed and attacking the basic operation of thought – that is, the capacity for linking. After all, a mental operation based on the pain-avoidance principle restricts the mind's capacity to take in the situation and constitutes a kind of partial or total anaesthesia, or self-exclusion of a mental function.

Let us return to our earlier example of sadness, considered in more general terms as a painful malaise of which we seek in some way to rid ourselves. The reasons for it may be real or existential, or it may be due to any number of internal and unconscious factors. The feeling we have is a message stemming from the operations performed by the mind on our internal situation. This process of affective semiosis (Fornari 1976) can, as stated earlier, be dealt with by the mind in two different ways. In the first case, the feeling is seen as constituting information about our interpersonal relations and internal processes; it can therefore undergo further processing, enabling us to seek the inner reasons for it and ultimately to act on and modify external reality – that is, to develop processes and behaviours directed toward solving the problem. The second form of mental operation, instead of attempting to explain the malaise and to establish its meaning, does its best to suppress it, so that it can no longer be felt. This means not that the inner state has been modified – in fact, it remains unchanged – but only that the subject can no longer be in contact with it, let alone experience it consciously. Some people always seem happy, content, euphoric and satisfied, at least on the surface, but on closer examination this state of apparent well-being proves to be due to the impossibility of accepting and processing the mind's internal signals of the unpleasurable feeling of sadness. And indeed, accepting these messages, becoming conscious of them, and supplementing this knowledge with an understanding of their inner reasons and of the conditions for modifying reality entails painful mental work, which such people cannot tolerate because it would arouse too much anxiety.

In order to satisfy one's needs by effective action on reality, the mind must have developed a tolerance of such unpleasurable states. A person who cannot endure depression will somehow banish such experiences and thereby deprive himself of the instruments for understanding the negative vicissitudes of his life, so that his situation will progressively worsen. Conversely, if the mental operations carried out are directed toward the modification mainly of internal reality – for which greater pain must be tolerated – the mind will become capable of enduring frustration and painful states, and will develop further mental operations that will increase its operational capacity, which it will structure in thought that can also be used to arrive at possible solutions. A person who lacks this tolerance acts self-destructively (autotomically) on his own mind, so that, when faced with the possibility of perceiving something unpleasurable, he deprives himself of the mental function whereby such experiences might be perceived. Such mental operations, governed by the unpleasure avoidance principle, give rise to mental autotomy or, at any rate, call mental functioning into question in the sense of facilitating incorrect operations (falsification, alteration, distortion, or confusion of codes). This may be a form of pathology no less severe for being concealed.

Since the most intense experience a neonate can have is unpleasure – that is, an unsatisfied need – it follows that it is the ability to tolerate this state that promotes his capacity to structure thought, or to develop mental operations appropriate for tackling his individual needs. If unpleasure cannot be tolerated, thought cannot arise. In order for thought to arise, frustration must be tolerated, at least partly.

On the other hand, it is precisely the ability to perform mental operations that increases the apparatus's capacity to tolerate frustration; for instance, the ability to recall a good breast is what enables the baby to wait and to recognize the real breast. However, this ability is anything but automatic; it is already the acquisition of a capacity to organize the inputs of afferences in such a way that they can be 'read' in relation to traces laid down earlier (Chapter 4). Hence the seeming paradox whereby the capacity to suffer allows a development of thought that in turn increases the capacity to tolerate frustration and to gain further knowledge of reality, through the increasing development of mental operations. Conversely, the establishment of excessive self-destructiveness will make it less possible to tolerate pain and less likely that further mental structures will be formed. In other words, two self-potentiating circuits may be postulated, one destructive and the other constructive.

Put differently, pain may be said to summon both life and death – but life summons more life, and death further death. On the level of the nascent life of the mind, however, death is disguised as non-suffering. Thus, nirvana casts its spell over man and takes away his life. Moreover, in the unconscious structures, this spell accompanies man throughout his existence, as both child and adult.

To return to our neonate – that is, to our problem of identifying how 'the mind comes into being' – we must consider first and foremost physical pain, the mentalization of which seems to rest on the foundation of neurobiological mechanisms. These underlie the capacity, which is to be constructed, for tolerating mental pain and processing it positively, and hence for the construction of further levels of operational mental capabilities. Our attempt to identify the successive levels of protomental operations will continue on this basis in the next chapter.

Chapter 6

The paranoid-schizoid metabolism

6.1 'Incorrect' and 'correct' mental operations

A central aspect of the Kleinian theory concerns a particular mental state described as the paranoid-schizoid situation or position,[1] which is characterized by a specific form of anxiety and specific defensive organizations. Segal (1964b: 127) defines it as 'the earliest phase of development [. . .] characterized by the relation to part objects, the prevalence of splitting in the ego and in the object and paranoid anxiety'. Meltzer points out, too, that it is considered to be a 'particular type of mental conflict that commences, let us say, around the third month of life and continues throughout the life of the individual' (1979a: 131, translated). Our own analysis is consistent with Meltzer's view, and identifies the dynamics described by the Kleinian school in terms of primary *mental operations*.

With the establishment of the first internal objects resulting from the first protomental operations, the process that will lead to the distinction between imagination and reality commences; it is a continuous dialectic between, on the one hand, the attempt to distinguish a nascent thought of the subject himself from reality and, on the other hand, the tendency to deny the existence of anything different from self (avoidance of unpleasure), which stands in the way of this differentiation between thought and reality. The infant must move on from a state in which he imagines the world to be totally coincident with himself to one in which he accepts that there is something different from self, having an autonomous presence of its own.

1 In analysing the development of Klein's thought, Segal (1979: 122) observes that the notion of a 'position' allows a new approach to the conception of mental development: 'The position is not comparable to a phase of development of the libido. True, the paranoid-schizoid position precedes the depressive position; nevertheless, the constant fluctuations between the two positions makes "position" a structural concept rather than a chronological one. The term "position" refers to a state of organization of the ego and describes characteristically conjoint phenomena: the state of the ego, the nature of the internal object relationships, the nature of anxiety and characteristic defences.'

The phenomena described by Klein (1946, 1963) can be reconceptualized as a dialectic between the paranoid-schizoid and depressive positions (Ps ↔ D, in Bion's notation), or a dialectic between processes of unpleasure avoidance and of knowledge of reality.

The two terms, 'paranoid-schizoid' and 'depressive', are drawn from the pathology of adult psychotics (schizophrenics, paranoics and depressives), because, in these states, it is possible to discern, in exclusive form and raised to a pathological pitch, mental dynamics which, accompanied by others, are 'physiological' in the infant mind. The analogies and differences between what happens in the nascent mind and the situation in adult pathological states will be considered in detail below; for the moment, we can sum up by saying that the schizoid state involves the splitting (*schizein* = *to split*) of representations that ought to be singular for the objective representation of a single reality into a plurality of internal objects, whereas the paranoid state is characterized by the massive projection, or attribution to something in the outside world, of what actually belongs to the subject himself.[2] Depression, on the other hand, involves intense internal suffering. This basic outline indicates how the mechanisms described can be observed in the neonatal mind, as, indeed, they have in part been described.

The paranoid-schizoid position may be said to define the situation of the mind when functioning with *mental operations* that lack realistic validity – that is, ones directed toward the abolition of unpleasure, which, in this sense, are *incorrect* in terms of the recognition of reality. In the depressive position, on the other hand, as will be shown below, the mind, by enduring pain, attempts to construct operations appropriate to the recognition of reality – that is, *mental operations* that can be defined as *correct*.

Klein (1935, 1940, 1945, 1948, 1952) defined as the depressive position the processes of recognition of reality whereby the subject has come to tolerate a situation in which something good is not always available and belongs not to the self but to the external world, coupled with the acceptance of his own dependence on that world. Since this recognition involves the development of processes truly appropriate for cognition, as pointed out first by Money-Kyrle

2 Reviewing the historical development of Klein's thought, Segal (1979: 113f.) writes: 'In the "Notes on Some Schizoid Mechanisms" she [Klein] clarifies her views on infantile development preceding the depressive position. It is a phase of part-object relationships, and it is dominated by a persecutory anxiety and schizoid mechanisms. Fairbairn had used the term "schizoid position" to describe the original, split, state of the primitive ego. Klein had emphasized the ideal and persecutory aspects of the early object relationship and called it originally "the paranoid position". In 1942 she introduced the term "paranoid-schizoid" to emphasize the co-existence of splitting and persecutory anxiety. To conceptualize her many observations and her views of the early development she uses as a theoretical framework Freud's theory of the life and death instincts.' In fact, however, she makes no use at all of Freud's energy-and-drive model (Imbasciati 1983b, c).

(1961, 1968) and later by the entire Bion school (cognition here implying not only a recognition of the reality of external objects, but also a certain capacity to be in contact with one's own experiences without excessive barriers, or 'lies'), our own theory will stress the appropriateness or otherwise of mental (and/or protomental) operations to cognition. For this reason, it seems justifiable to label as 'correct' rather than 'incorrect' operations that lead to a more or less accurate recognition of reality, rather than to its projective or hallucinatory distortion.

The two positions can be regarded as homologous to mental functioning, as based respectively on the unpleasure-avoidance principle and on the reality principle: their ongoing dialectical elaboration enriches the mind in so far as the depressive dynamic prevails. Although they can be considered as two different types of object relationship (Meltzer 1979a, Bion, F. 1984), and hence described in terms of feelings, their working can nevertheless usefully be investigated from the point of view of the *logical operations* called for by the two positions. On this basis, we shall attempt to analyse the paranoid-schizoid position in terms of its typical mental operations and the respects in which they are 'incorrect' – that is, do not match reality. Let us take the example of the feeding situation.

In terms of the above description, a baby, under the stimulus of hunger, tries to summon up the memory of the breast and to hallucinate it, performing the operation of banishing the evil spirit that has taken shape inside him. The exorcistic operation involves a hallucinatory memory – that is, the evocation of a memory trace – which, however, is not recognized as such; that is to say, the neonate cannot yet distinguish it as a true memory, as referring to something that has existed in the past outside himself. Instead, it is experienced as produced by himself and as if it actually existed in a total reality in which external and internal reality are merged. This is therefore a hallucination constructed on the basis of what could have been a memory; for convenience, we shall call it *HM*.[3]

Let us, then, continue our schematization of these protomental dynamics in terms of logical operations. The exorcistic operation (*E*) may be deployed in two cases: (1) when *HM* is used in order not to feel the hunger – that is, to attenuate the signal of hunger that is about to arise in the mental apparatus, replacing it with another, self-produced, signal that does not correspond to reality; (2) when the hunger is satisfied because food happens to be given to the child, whose mind experiences the disappearance of the pain of hunger as resulting from its own exorcism.

The afferences that could be read as a set capable of becoming perceptual, and might be referable to previous sets that could have become established as

3 The hallucinated memory is here different from a hallucination experienced as remembered.

memories (potentially serving for perceptual recognition), are confused with and superimposed on traces that take the form of hallucination. If, then, the afferences that could give rise to perception are denoted by *PM* (perceptual memory), in the exorcistic operation, *PM* tends to be equated with *HM* ($PM \rightarrow HM$); in other words, what could be organized as perception, or recognition of an object existing in an external reality,[4] is equated with the hallucination. Such an operation is obviously 'incorrect'.

Conversely, the exorcism fails if the baby is not fed and the hunger becomes too intense to be masked by the hallucination. If *E*+ and *E*– denote the success and failure of the exorcistic mental operation respectively, the various situations can be illustrated by the following diagram:

E	+	Hunger masked Hunger satisfied	*HM* *PM*	$PM \rightarrow HM$
	–	Persistent hunger	*HM*	*HM*–

This diagram conceptualizes the two situations in logical terms and illustrates their contradictions. These contradictions constitute the typical difficulty faced by the neonatal mind before it becomes able to perform a mental operation valid in terms of reality: *E*+ involves a $PM \rightarrow HM$, and hence confusion, as well as an *HM* experienced as if it were effective and hence having a positive sign (good), while *E*– also involves an *HM*, but of opposite sign. Both of these result from the same reality situation, the absence of the breast, but are mutually contradictory; discrimination is therefore lacking in the operation. Moreover, the situation is further complicated by the fact that *E*+ may stem from two fundamentally different reality situations – namely, hallucination on the one hand and an event that ought to become perceptual on the other hand. The positive sign depends not on the presence of the object in reality, but on the fact that the baby has become less sensitive to hunger owing to a mechanism of illusory satisfaction. In order to accomplish a realistic – that is, 'correct' – operation, which thus distinguishes between recognition of reality and simple memory (*M*) or hallucinatory memory (*HM*), the mind must differentiate between *HM* and the event that

4 This object may also simply be hunger – that is, the perception of an objectively existing hunger.

can become perceptual (*PM*). This discrimination proves to be very difficult, both on account of the contradictions mentioned above, and because an *HM* → *PM* would signify the prospect of something good that is not available and does not belong to the incipient self, which attempts instead to create something to banish the monster of hunger. This would give rise to a completely new situation for the nascent apparatus, describable, in somewhat adultistic terms, as a primary feeling of chagrin and rage, leading to rejection; in this case, the rejection would be of an operation, *PM*, experienced – wrongly – as useless. In other words, the nascent perceptual thought would tend to be rejected.

The above description of the case of hunger could be adapted to represent any other situation in which a signal biologically predetermined for the signification of an unpleasure (pain or danger) gives rise to a relieving response involving the provision of external help to the child, rather than an anaesthetic, hallucinatory mechanism. Further complication may of course be introduced by the intervention, not only of the external world, but also of internal 'succour' resulting not so much from exorcistic operations as from already consolidated 'good memories' stemming from fetal experience. Conversely, the process may be determined by prior 'bad' fetal experiences. Hence the importance of fetal experience in laying the foundations of the future mental structures. In this field, however, much remains to be investigated: what experiences, or rather what modulations of the fetal experience mediated by the dialogue between the expectant mother and the baby in her womb, may lay the foundations for good, subsequent post-natal development or, conversely, for unfavourable development? Through what assemblies of afferences and what maternal modulations of these assemblies might the qualities of 'good' and 'bad' already become established in the fetus? And how do these come to constitute a good or bad prognosis for the neonate's mental development?

6.2 Schizoid mechanisms and the birth of perception

Despite the initial tendencies for the hallucinatory mode to be reinforced, the proto-mind can, it seems, acquire the capacity to distinguish between *HM* and *PM*. This possibility can be attributed to a certain equilibrium between the three ambiguous combinations illustrated in the above diagram, in terms of the frequency of the prevalence of a certain event. If the baby's hunger is satisfied before he becomes aware of his need, he is unlikely to be able to distinguish between the hallucinatory and the perceptual memory; in order for this to be possible, a certain time must elapse between stimulus and satisfaction, during which the baby is able to summon up something intended to get rid of the sensation of hunger (*HM*) and to experience its failure before receiving the actual satisfaction of the bottle, so that he has an opportunity of perceiving it as such (*PM*) and distinguishing it from *HM*. The absence of this optimum interval will constitute an obstacle to distinguishing between the event capable of becoming perceptual (*PM*) and the hallucinatory memory (*HM*). If *PM* is prevalent prior to the accumulation of a sufficient *HM*, it is less likely that the two will be compared and that *PM* will therefore be read as such (perception). Furthermore, a dearth of *HM* impairs the functioning, inherent in the surrogate satisfaction, of operational units serving to admit a satisfaction-affording event and then possibly to distinguish it from others that also signal a forthcoming satisfaction (*PM*) in reality.

However, if the interval is too long, the first mental operation constituting the aggregation of a hallucinatory memory becomes meaningless; the baby's state can in this case be imagined as negative, as made up of confusion, despair and rejection of everything that had previously been constructed in his mind. If the delay is excessive, there will be a permanent conflict between the positive and negative hallucinatory memories: one and the same hallucination is split into the signalling of two events of opposite sign. In other words, there will be a conflict, in that one and the same operation, *HM*, although carried out with the same exorcistic purpose, sometimes succeeds and sometimes fails, giving rise to a perhaps rage-like situation of confusion between *E*+ and *E*–, between a successful and an unsuccessful (hallucinatory) 'thought'; that is to say, conditions favouring abandonment of the first mental anlage may arise, as if the conclusion had been reached that 'thinking is no use'.

Conversely, an optimum interval may be supposed to favour the possibility of differentiation, in the functioning of the primitive mnemic aggregate, of two mental sets with two different functions, one of them recognizing the hallucinatory memory as such, and the other capable of recognizing the memory when confirmed in a perceptual event.[5] In this way, the baby can come to understand that the hallucinatory memory has an illusory function – that is, that the positive hallucinatory memory is in reality negative – whereas satisfaction comes from *PM*. In other words, if

$$HM \rightarrow PM = \text{satisfaction},$$

it follows that

$$HM \neq PM,$$

and, further, if HM_{E+} denotes the memory trace when the hallucinatory exorcism appears to have succeeded,

5 The optimum interval cannot be quantified as a single value applicable to all neonates, but depends on a large number of variables that make up each individual mother–baby relationship.

HM_{E+} = illusory if there is no PM.

In other words, the baby becomes able to distinguish between the hallucinatory set HM and the recognition of a perceptual event – that is, he can separate hallucination from perception – and this potentially leads to the conclusion that the hallucinatory exorcism that was seemingly successful actually succeeds only when a PM is recognized; otherwise it is illusory. In order for these probabilities of distinction to be possible, however, there must be enough PM, to be compared with a 'sufficient' HM, as HM_{E+} and HM_{E-} will otherwise conflict.

If, therefore,

$$HM \neq PM,$$

it follows that

$$H \neq P$$

– in other words, an incipient distinction is being made between perceptual and hallucinatory experiences. It also follows that

$$\overline{M} \neq H$$

– that is to say, in a memory trace that is always the same (M), the difference between the occurrence of a perceptual event and that of a hallucinatory event is recognized, so that it becomes possible to recognize the difference between the recurrence of a trace experienced as hallucination and the emergence of a trace experienced as a true memory ($\overline{M}$) – that is, a trace experienced as the representation of a real external object that is not present. Of course, the 'memory of an object' or its 'perception' by no means signifies that the external object is remembered, represented or perceived in the infant mind as it is in reality; both infantile perception and infantile memory exhibit an extremely high degree of apperceptive distortion (Imbasciati 1967), which means that the primitive internal objects are representations that never correspond to external objects. What is important at protomental level is the establishment of the correspondence between signifiers and signifieds whereby an internal object can be distinguished as a memory rather than as the perceptual recognition of a certain assembly of afferences, which, precisely on account of this type of recognition, is perceived as something external and present. In other words, it is a matter of the experience that something internal, emerging in the mind, corresponds, in certain very distinct cases, to an object that is present and, in others, to a past experience of objects that *have been* present.

If, then, the set constituted by the hallucinated memory is distinguished from that of the perceptual memory, the various components too are distinguished. An incipient distinction is made between the process of hallucination and that of perception; however, by a further step, it is also possible to differentiate the memory, identified with something referring to an object that has been but is no longer present, from the other type of engram, constituted by *HM* – that is, from something experienced as self-produced and at the same time real. In this case, the memory can be used to wait for the perceptual event. In other words, at this point the trace is usable for realistic rather than for hallucinatory and exorcistic operations. This is a fundamental step, because it constitutes the foundation of the gradually acquired capacity to be prepared to 'grasp' – that is, to receive – the object. The perceptual capacities can then develop, laying the foundations for 'effective action' – action that takes account of the perception of objects as they are and therefore involves an actual approach to the external object, such as stretching out the hand to the bottle. At the same time, the appearance of a *PM* facilitates the development of *HM* (experienced as self-produced) out of an *H*, experienced as perceived – that is, as if it were real – but no longer self-produced, whereas, on the other hand, the experience of a self-producing something causes *HM* to shift toward an experience of imagination – that is, of something self-produced but non-existent. A metabolism of this kind is presumably active in an adult too, in the ongoing transformation of memories, mediated not only by the succession of perceptions and consequent memories that modify their predecessors, but also by the introduction of an activity of imagination 'into' the operations of memory transformation.

The birth of the possibility of perception can be regarded as the *fifth level* of protomental functioning: the subsequent steps may be regarded as 'mental' rather than protomental. On this fifth level, the capacity for perception lays the foundations for a differentiation of the 'other-than-self': whereas the system could previously conceive only of an initial anlage of self as opposed to a 'not self', the 'not self' now takes on a positive value of its own – namely, that of reality, perceived as such, which therefore forms the basis of the 'other-than-self'. This has as its counterpart the construction of a self as an interpreting, percipient subject.

A further function of a memory distinguished as such can also be described: it is the use of the memory itself for the purpose of waiting for and recognizing the real event, and hence as an initial anlage of a representation of the *absent object*. The immediate consequence of the acquisition of perception is therefore the capacity to perceive objects that are present in reality and, immediately afterward, the capacity to distinguish these from possible absent, 'remembered',[6] objects, which, however, are considered to be real and distinct from the hallucination. Hallucination can now (see Section 4.3) in

6 It is again emphasized that this 'remembering' has nothing to do with consciousness, which is exhibited only, and then not always, in adult remembering.

turn be distinguished from a new category of experiences – namely, imagination.

The baby remembers that the bottle exists, can think that it will come, and will therefore be able to endure hunger without any longer masking it. At the moment when he first recognizes a memory as such, he begins to acquire a first idea of time, and at the same time of an object that may not be present; this object may in turn be considered to be real, and existing elsewhere in time and space, or experienced as 'imagined'. In terms of feeling, the toleration of hunger while waiting to be fed may be said to constitute the mental foundation of hope and the trust that an expected event can become a reality.[7] The trace is therefore used no longer to banish hunger, but to endure it, and this is the child's first realistic mental operation. The initial pattern now undergoes the following modifications:

$E \to O$	+	Hunger endured Waiting	$\longrightarrow \overline{M} \to (\overline{M} \to EV) = P+$
		Hunger satisfied	$\longrightarrow PM \to P \to P+$
	–	*HM*	$\longrightarrow H \to H-$

The exorcistic operation, which, during this phase, takes on a degree of efficacy ($E \to O$) by virtue of a certain correspondence with reality,[8] gives rise to two further events – of success and failure – but now, in the case of the positive event, there is no longer a hallucinatory memory (*HM*), but a true memory ($\overline{M}$), which, when coupled with the recognition of the possibility of an external event (*EV*), with the expectation of and the search for that event, gives rise to an increase in perceptual capacity (*P*+).[9] Hence, the formation of

7 This may constitute the first step on the way from need to 'wish'.

8 The notation $E \to O$ indicates that the exorcism is now no longer the rejection of something, but the search for an 'object' – in other words, that the exorcistic operation is capable of becoming an operation of objective recognition of reality.

9 Here we might discern the beginnings of curiosity, as typically manifested in young babies after the second month of life.

memory leads to the enduring of hunger and to waiting, and facilitates the possibility of recognizing the external breast, the bottle, and then other external objects; in addition, the foundations are laid for searching for such objects or for a protocommunication that signals to the mother that the baby is searching for something. The intervention of the real breast thus gives rise to a *PM* that quickly becomes *P*; indeed, to the extent that there is a waiting period endured, it will become a *P*+ or, in other words, an increase in the child's perceptual capacities. Since there is no longer any ambiguity, the exorcistic operation becomes realistic. Conversely, should the state of non-satisfaction persist – that is, should the successful exorcistic operations ($E \rightarrow O+$) and subsequent ones fail, there will still be an *HM*, which, however, because $E \rightarrow O-$, may be experienced negatively. For this reason, what is now more properly experienced as hallucinatory (*H*) becomes susceptible of abandonment by the mind (*H*–), or may undergo a shift toward the construction of the imagined experience.

Let us now continue our schematization of the successive levels of mental and protomental operations. To recapitulate, the first level constitutes the appearance of the first bad proto-object (Chapter 4); the second is that of autotomic hallucinatory exorcism (Section 5.1); the third is the formation of a first 'good' memory trace, which is hallucinatory (Section 5.2); the fourth is the stabilization of the categories 'inside = good = self' and 'outside = bad = not self' (Section 5.3); and the fifth is the birth of perception (this section). We come now to the *sixth level*, just described, in which a certain – balanced – exercise of hallucination and painful experience, in the feeding relationship or other early interactions, leads not only to the first capacity to perceive reality, but also to the capacity, which undergoes particular development in the human species, to remember objects (or events, or even one's own states when not current), to perceive time and space, to have trust and hope, and, finally, to 'imagine'. In other words, the distinctions illustrated in the diagram in Section 4.3 are attained.

Of course, the progression of levels does not mean that, once a given new level is constructed, the previous one ceases to exist – for the system may return to more primitive levels of functioning. In particular, the fifth level is characterized by continuous alternation between hallucination and perception; although the 'birth' of perception can be placed on this fifth level, its development and stabilization are the prerogative of the sixth: whereas the paranoid-schizoid situation involves hallucinatory mechanisms (the attribution of internal states to the external world), its exercise allows a different use to be made of it, tending toward the mechanism of perception. The Ps $\leftrightarrow$ D oscillation, recognized since Bion as constantly present in the formation of 'thoughts', tellingly expresses the way in which the hallucinatory mechanisms of Ps can, by repeated exercise in a wide variety of communicational situations, lead to D. In our conceptualization, therefore, the alternation between paranoid-schizoid and depressive mechanisms is 'explained' as the possibility of parallel and reversible operations involving different assemblies of memory traces.

6.3 'Envy of the breast': proto-envy as an attack on thought

We have shown *schizoid*-type mental operations to be characterized by the fact that one and the same real external object gives rise to different, separate (that is, split) internal images, which can be partly eliminated, or rather evacuated, by autotomy of the mind's incipient functioning; and we have also shown how, in these operations, there is no distinction between internal and external, or between hallucinatory memory and the operation of perception. Thought, in these processes, is hallucinatory in nature, and determined by a fundamental and inevitable autotomic tendency of the mind. The *paranoid* aspect refers more properly to the expulsion of the bad parts and the attack on these parts when projectively identified in external objects. This aspect was described by Klein (1957) as an envious attack on the breast.

Envy may be said to be man's most primitive and stubborn feeling,[10] which is thus identified with hate, and can be seen as the wish to destroy what appears good. Again according to Klein (1957), it differs from jealousy, which is based on love and on defence of a privileged link within a triangular interpersonal situation. Envy, for its part, arises in the context of a dyadic relationship and tends toward the appropriation and/or destruction of desirable qualities in the other. The popular tendency to confuse these two emotions can be interpreted as an attempt to disavow the feeling of envy, which is felt to be more shameful than jealousy.

The infantile roots of envy were identified by Klein (1932, 1957, 1963) in the relationship between the neonate and the first object worthy of appreciation – namely, the breast. The envious attack is described as a series of 'fantasies'; that is to say, it is reflected in a succession of 'images' (which are of course unconscious) that use material from the baby's experience: emptying the breast of everything valuable, annihilating it, swallowing it and causing it to disappear, and later tearing it to pieces with the teeth. Later still, as the child gains experience of the world and constructs his own mental processes by the use of other parts of the body – such as the eyes, hands

10 Envy is preserved as the human mind's most stubborn, hidden and active feeling. From its origins in relation to the breast, it is elaborated into envy of everything that is offered and everything life-creating. In particular, it arises in connection with the maternal functions and the female functions in general, underlying both the development of femininity and male feelings toward women. In every culture, woman remains the most powerful symbol of creativity, for she is Gynè, the creature that creates (Imbasciati 1979a).

or sphincters – he will work with the traces of the relevant afferences, so that his fantasies will be describable in terms of images of the action of these organs. As the mind gradually develops, the fantasies concerning the breast, which is represented in more and more complex form – as the mother's body, as the mother, and as a person – will become increasingly cruel and sadistic.

The Kleinian conception of powerful aggressive fantasies in newborn babies has been much criticized. Neonatal observation seems not to bear it out, at least in the majority of cases. Our own conceptualization in terms of variable assemblies of memory traces accounts for both the Kleinian theses and the critiques directed at them. The critiques are confirmed by the fact that the idea of the existence in a neonate of aggressive fantasies, and still more of envy, involves the attribution of a much more advanced mental level to him. Our cognitive logical schema attempts a different approach, in which the construction of the mental operations conceptualized as an attack on the breast, or on the first object, will be described in cognitive terms, as an attack on thought. After all, it is possible to describe protomental operations that constitute this attack and to regard them as precursor events of what can, later, more properly be called envy: we may adduce a proto-envy, and redefine what Klein adultomorphically described as an aggressive fantasy in the neonate as a protomental operation with destructive effects on the nascent capacity to think.

As stated earlier, an infant tends to experience everything good as belonging to himself, so that the self is established as essentially good and opposed to the not-self, which is referable to everything bad. When he succeeds in formulating the first operations that conform in some way to reality, he identifies something good that is no longer self, but an external presence (perception: the fifth level). From the opposition of self and not-self that prevailed in the first days of life, he moves on to an opposition between self and an external world that could also be termed 'breast'. This follows from the establishment of the first perceptual processes: if

$$(\overline{M} + EV) = P+,$$

then

$$P \neq \text{self, and also } P = \text{good.}$$

Hence:

$$\text{self} \neq \text{good}$$

$$\text{self} = \text{bad.}$$

To the extent that an external 'good' that is other than self presents itself, what has been experienced as bad and evacuated so as to constitute the external monster – or evil spirit – to be exorcized, tends to be reassimilated as 'something' occurring within the subject himself and therefore coinciding with the self. This possibility, for a mental proto-apparatus still functioning in accordance with the pain-avoidance principle, is intolerable, so that this 'something' must somehow be split off again from the incipient self. Since there is a shift from a non-self experienced as the absence of self, except in so far as it is the locus of appearance of the 'monster', to a not-self that could be a presence and a good presence (in other words, the experience of an 'external world' and the possibility that this world might consist of a 'good breast' is contemplated), the fact that a bad experience is felt to belong to the self is certainly a contradictory event. The possibility of feeling a part of the self to be bad becomes a token of the constitution of an intolerable, bad internal object. This was already preformed from the aggregation of afferences of hunger, but had been promptly evacuated in primitive projection. Subsequent experiences, following the event of perceptual processes, consolidate and give a more appropriate physiognomy to the first bad object, which thus tends to be forcibly reintroduced into the self, giving rise to something definable as suffering, fear and rage. In opening itself to reality, the neonatal mind also opens itself to the perception of something internal that is bad and unpleasurable, in contraposition to a good object. The roots of that object lie in the first true perceptual memory, which, as we have seen, serves the purpose of waiting, of enduring hunger, and of initiating the operations that can result in the reappearance of the wished-for object. However, this memory is at first weaker than the bad internal object, because hunger and other experiences of need take the form of bad objects that the mind is unable to offset by appropriate realistic operations, and because, moreover, these are not recognized as belonging to the self and hence experienced as good instruments or good constituent parts of the self.

The system now attempts to bring the good back into itself and to expel the bad. It may once again try hallucinatory mechanisms, but these may prove inappropriate if, as in the normal situation, the perceptual processes have prevailed over hallucinations in spite of the difficulties. The situation that then arises can be described as pain, chagrin and rage; and in it may be discerned the root of the envious attack on the good external object that does not allow itself to be internalized, or owned. However, it is a 'root', conceivable in information-processing and constructivist terms not as an aggressive fantasy, and still less as a feeling of envy. The painful situation intensifies the experience of something unpleasurable – of the bad within the self – and therefore tends to induce a further projection of this additional bad object. This projection takes the form of an attack on the good external object, which is therefore experienced projectively as attacking and therefore bad; by this process, it is not only attacked, but also, as it were, corrupted in its original goodness. In terms of mnemic assemblies, this object, which had been capable of being perceived as external – that is, by the use of an appropriate engram – is once again read by a less appropriate engram; this is what we mean by the 'corruption' of its goodness. For this reason, if envy is definable as an attack on the good, in our case it is an attack on an object that has, through paranoid projection, become subjectively bad. Such an attack is now also perpetrated on the infant's own nascent mental processes, which are felt to be inadequate and are therefore hated, and experienced as bad instead of good attempts. The result is an attack on the subject's proto-thoughts, which immediately becomes an attack on the first reality-matching perceptions; in this sense, it is an attack on the good external breast. This constitutes the *seventh level* in our schematization, involving what has been defined as paranoid projection, with the consequence of persecution.

6.4 Paranoid circuits

The child is increasingly exposed to an intensification of the experience of a bad internal breast, its consequent projection and hence an attack on the external breast, which is an attack on the perception of reality, on thought, on the goodness of the instruments the mind was in the process of creating. However, there is in addition a situation that opens the way for the infantile mind also to feel something good inside itself, and hence also to reinforce the memory of a good breast that has been experienced, to consolidate its memory traces and to strengthen its perceptual capacities – namely, the situation that follows the use of the feeding experience. This can be described in terms of the preservation and stabilization of mnemic aggregates that can be used for certain mental operations. A neonate has inside himself a proto-image of the external breast, an internal breast (B_i) with the potential to reproduce and represent the external breast (B_e) and its goodness, but since this external breast is not always available, this B_i that was capable of making B_e a present reality becomes a mere memory M, which can readily be submerged by pain and distress, or erased as a trace in such a way as to compromise recognition (P) of the world. This internal breast needs to be reinforced in order to be accepted as such, in order for the trace to be preserved as a good representation of something good which, however, is not available. This reinforcement may occur during feeding.

While being fed, a neonate feels that he can equate the external breast (B_e) with the internal breast (B_i). This experience is in part illusory, because the baby cannot have for himself the 'good' that is given to him by a reality external to himself and that he experiences as created by himself (Winnicott 1958, 1971). However, it also possesses certain characteristics of reality, because at that moment B_i allows him to 'perceive' B_e: at that moment, he is actually sharing in the reality of the offer of the breast by perceiving it – that is, by accepting within himself the object that is coming to be recognized as indispensable, but not yet as belonging to another. The breast is accepted and is 'inside' him, not only physically but also mentally; in this situation, in so far as the baby derives satisfaction from the relationship with the breast, so, too, the realistic mental processes are reinforced and stabilized. Hence, there is not only the physical possession of a breast constituted by the bottle and the milk, but also the mental possession of an internal 'breast' represented by the reinforcement of the processes that have proved to be operationally effective: this is the capacity to 'perceive' B_e on the basis of the recognition made possible by the stabilization of the trace B_i. Hence:

$$B_e = B_i$$

However, if there is hunger without real satisfaction of the need, an experience of profound difference between the internal breast (B_i) and the external breast (B_e) becomes established:

$$B_i \neq B_e$$

The external breast is experienced, as it were, in hyperbolic terms, as infinitely large and rich. The internal breast (B_i) is experienced as a memory, but endowed with little intensity, or, in operational terms, it is relatively useless compared with the perceptual appearance of the external breast. So, if B_e is taken to be infinite, B_i in memory tends toward zero. Hence:

$$\left. B_e = \infty \right)\left(B_i = M \to 0 \right.$$

This difference, which may be equivalent to what has been described as greed and is itself very painful, is therefore projected, and it, too, like envy, enters into the metabolism of the paranoid circuits: the situation of comparison between B_e and B_i is unbearable, so that the system seeks to erase it by attempting to revert to the old operations of bringing the good inside the self and expelling the bad. While perceptual capacity is reduced, so, too, is the experience of the external breast (B_e–) by an operation describable in somewhat adultistic terms as a feeling of attacking the breast, and in operational terms as an attack on the baby's own mnemic-perceptual capacities. At the same time, what the child feels he possesses assumes gigantic proportions – that is, the image of the internal breast (B_i+), but of a B_i no longer capable of being equivalent to B_e, or of operating through perception. This can be expressed as the hallucinatory transformation of memory ($M \to HM$): the illusion arises that the memory of the breast is a breast possessed as a good entity in reality.

However, such an operation has already been rendered falsifiable by the child's experience: realistic-type operations create a precedent inconsistent with the attempt to reinstate hallucinatory thought, thus giving rise to a situation of contradiction and confusion, which is experienced subjectively as a state comparable to that of extreme rage. The orientation of the above operations in one direction rather than its opposite has been described in terms of oral experience; obviously, however, the presence of prior engrams can play a part in the modulation of the system's nascent operational capacities in one direction or the other. Conversely, prior or concomitant negative engrams (such as physical pain) will give rise to a contrary modulation.

The affective states described in the clinical tradition as fundamental to early infantile development are envy and greed; we have endeavoured to translate these into logical and mathematical terms, as the *metabolization* of mental sets derived from traces of afferences, and have essentially described a proto-envy, with its connotation of a destructive (and self-destructive) tendency. The infant mind is unable to tolerate a 'bad' experience within itself, and therefore tries to eliminate it from itself by projecting it outside; it projects it into the same external 'breast', which thus takes on all its bad, 'envious' and attacking characteristics. In this way there forms an image of the breast-as-external-world that persecutes the child's mind, because it has taken in the rage and destruction that the mind has found itself to have inside and that it has been unable to tolerate as belonging to itself; the result is the formation of a mnemic-imaginative set describable as a persecutory breast (B_p). The counter-attack, also describable in terms of a precursor of envy, is unleashed against this persecutory breast, which is experienced as external, but is in reality made up of parts of the child's own self.

With regard to greed, the attempt to reinforce the subject's own internal breast (B_i+) can thus be interpreted in the same way as the wish to possess the real breast and to take it into the self. This operation, too, because it is not effective, is experienced as 'bad' and is therefore rejected to the outside. The experience of hunger now takes the form of a bad object that is projected and contributes to the formation of an even more gigantic breast.

To sum up, the result of the so-called greedy and envious attack on the breast, whose aim is to reduce confusion, is a projection to the outside of 'bad' proto-feelings that cannot be recognized as belonging to the self. In this way, there comes into being an envious, greedy, hungry external breast, which is 'bad' because it is the container of the rage, aggression, greed and hunger, and is experienced by the baby as an object that can in turn attack him, angrily devouring him with its mouth. The more the child's hunger increases, the more the breast threatens to eat him up and annihilate him. The result is an anti-realistic feedback circuit based on two specific mental operations, one of which is the abolition of the difference between inside and outside, and the other the attribution of what is primarily internal to the outside. At this point, the external breast (B_e) has become the child's persecutor. The emotional dynamic described here can be represented by the diagram (Figure 6.1), using the same notation as before for the correspond-

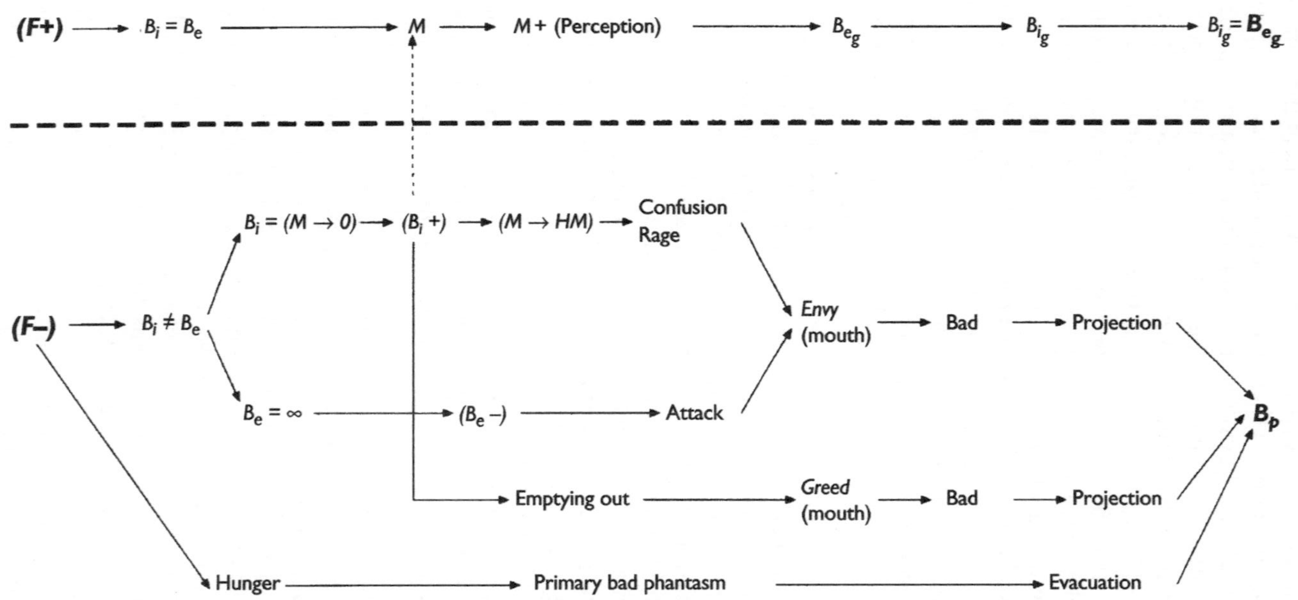

Figure 6.1 Two possible chains of protomental operations, one (above) leading to perceptual recognition of reality and the other (below) to the formation of the fantasy of the persecutory breast (see text for detailed explanation).

ing traces or mnemic-imaginative configurations, or operational units of proto-signifiers.

Figure 6.1 illustrates the two situations arising in the presence and in the absence of feeding: they are denoted by the symbols *F*+ and *F*– respectively (*F* = feeding). The intensity of B_p (persecutory breast) follows from the fact that it derives from three concurrent mechanisms, of which the first involves envy, the second concerns the need to increase B_i (that is, greed), and the third is the primary mechanism of evacuation stemming directly from hunger. Through projection (*Pro*), all three converge in the generation of B_p. With regard to situation *F*+, one possible theoretical scenario is a partial operation of recovery of B_{i_g} stemming from the persecutory mechanisms: the need to potentiate B_i might be reflected in the tendency to offset the erasure of the memory by a countervailing attempt to stabilize it.[11] This possibility is represented in Figure 6.1 by the dashed vertical arrow extending upward from B_i+ to *M*.

The operations we have attempted to identify thus give rise to a persecutory breast experienced as external, greedy, envious, attacking, angry and, since these modalities are fantasized in oral terms, devouring. The fear of being eaten up persists for a long time in children's minds. At first, this persecutory breast is stronger than the good image of the external breast. This can be illustrated by the following notation:

$$B_p = B_{e_b} \text{ and } B_p > B_{e_g}$$

At this point, a self-potentiating paranoid circuit arises in the infant mind, leading to the possibility of ever higher persecutory intensity. According to the classical psychoanalytic description, the presence of a persecutory breast results in an upsurge in fear and in a tendency to counter-attack, so that these 'bad' feelings are in turn reprojected to the outside, thus increasing the terrifying nature of the persecutor and hence the need for a more intense counter-attack. This potentiation of the persecutor can be described as follows in our cognitive terms. The mind has 'learned' that operations involving memory – $f(M)$ – serve to refine perception and hence contact with the external world, and also make it possible to have the real external breast; in other words, they serve somehow to 'take in' (in fantasy, with the mouth) the external world and breast, so that these remain 'in the mind' as good memories, constituting the foundation of the capacity for good mental operations and capable of ultimately being consolidated in a good internal presence (B_{i_g}). Presumably, therefore, the mind is led to apply this operation of 'taking in'

11 The operations leading to consolidation of the good internal breast could also be described in terms of the 'subjective object' and the 'creative illusion' (Winnicott 1958, 1971).

also to the persecutory external breast that is in the process of formation. The result is the reintrojection of the bad breast. Put differently, the establishment of perceptual processes, and hence of appropriate representational processes, involves an external experience which, however bad, has a homologous internal counterpart.

But in this way, the presence of a bad internal breast (B_{i_b}) is reinforced, giving rise to further pain, fear and projective mechanisms (Pro), which increase the degree of terror induced by B_{e_p}. This is therefore intensified, resulting in a further increase in the intensity of projective mechanisms, and, in turn, of the intensity of persecution. The result is a vicious circle, which can be illustrated in diagrammatic form as follows:

$$f(M) + B_{e_p} = B_{i_b} \rightarrow fear,\ pain \rightarrow Pro \rightarrow B_p \rightarrow (B_p)^n$$

The paranoid circuits therefore tend to be self-fuelling. This means that the neonatal mind experiences a very difficult situation, which may justly be compared to the persecutory delusions of adult psychoses, by which it risks being overwhelmed. The possibility of escaping from this persecutory metabolization of the first fantasies will depend on the extent to which the paranoid circuit can be interrupted by the quantum of B_{i_g} that has been able to form. The symbol B_{i_g} denotes not only the elaboration of oral experience as described above, but also any other (neonatal or fetal) elaboration resulting in an increased capacity to recognize the 'good' in a reality perceived as external. This capacity in turn helps the subject to seek other external sources of 'good' and to recognize and reintroject them, so that he can also contain within himself the bad internal breast without projecting it and hence without increasing the level of persecution. This will facilitate reinforcement of mental operations suitable for the recognition of the external and internal worlds, with the relevant representations of the external world and the relevant functions of relating to it. If, conversely, the persecutory breast is reinforced, the mind will remain occupied in counter-attacking it – in a sense justifiably – and will continue to engage in destructive operations antithetical to cognitive operations.

6.5 Projective splitting and perception: the percipient self

The analysis of the paranoid-schizoid position in Section 6.2 shows that recognition of the perceptual event (P = percept) signifies recognition of something outside and different from the subject's self, and constitutes the first realistic, or effective, operation ($P \neq$ self). The external other-than-self (P) thus takes on a character describable as affective, as it is recognized as something 'good': what is perceived is sought after. This implies that the self can be experienced as a negative entity, since the painful state of hunger is now identified as belonging to the sphere of the self. The other-than-self now comes to assume the form of something good and the self that of the bearer of suffering (which is bad); this is the antithesis of the first situation organized by the mind to avoid unpleasure. All the previous organization is based on the mechanism of *projective splitting*. This is the process whereby what is in reality generated by the individual is separated (split off) from the self and then attributed to something external. Logically, splitting precedes projection: in order for there to be projection, the external must be characterized as in contraposition to the self and not merely as something remote and vague: at the beginning of life, the alien is pure absence.

On the basis of the organization underlain by projective splitting, the mind splits off from itself every unpleasurable event, which it experiences as originating from the outside world – that is, it projects them on to the outside world – whereas, conversely, events stemming in reality from external objects are, if pleasurable, experienced as born autochthonously of an omnipotent and hallucinatory self. Acquisition of the possibility of organizing perceptual processes and retaining memory traces experienced as such entails a reversal of the entire prior organization. This is the threshold of the *eighth level* of protomental operations. The problem now arises of how the mind can reconcile the new nascent outlook with the previous structure. With the consolidation of the possibility of perceptual and mnemic operations, comes the recognition of an external world that is no longer the mere negation of self, but is the bearer of positive characteristics. This implies the possibility of the existence of a self endowed with 'bad' qualities. Given the possibility that the external world too may be composed of good objects, the mind likewise continues to use its mechanisms, moving on from predominantly schizoid processes to a paranoid-schizoid situation when the external world is massively and continuously filled with projections of internal destructive states. In this phase, the external world is experienced simultaneously as good and bad, so that, in parallel with the progress in the recognition of the external world as present and good independently of the child's need, there persists the experience of living in a world also pervaded by cruel monsters, created by the projection of anything that is found to be bad in the self and is expelled (the projection of suffering, rage and any pain).

Hence, the situation illustrated in the upper part of Figure 6.1 represents a mode of mental functioning – the mode that leads to M and B_{i_g} – that is becoming consolidated simultaneously with the exacerbation of the projective events (portrayed in the lower part of the diagram); indeed, this mode of functioning comes into being in so far as phenomena of projection proper – that is, paranoid-schizoid phenomena – are superimposed on the primary evacuatory phenomena of projective splitting, which are merely schizoid. The paranoid-schizoid phenomena are due to the projection of every increase in bad internal events (B_i+), which converge in the ultimate formation of a persecutory external world (B_p). The upper part of Figure 6.1 is separated from the lower part by a dashed line, which indicates that the two situations are simultaneous not only by virtue of the coincidence of the alternate vicissitudes of feeding ($F+$ and $F-$), and are not depicted together solely for the sake of clarity of exposition, but are present together and interconnected. Their simultaneous presence is due mainly to the fact that the external world is becoming, no longer the absence of self or of its parts, but the presence both of good objects and, at first to a greater degree, of projected bad objects.

The attempt to recover for the self the good recognized as being outside and to take it inside is thus superimposed on the mechanism of projective splitting off of the bad parts. The child would like once again to see the external breast as its own, reverting to the first operation performed in the hallucinatory mode, but now this is achieved not by simple autoplastic mechanisms but by considering the presence of an external reality, believed capable of being made to belong to the self. Since this operation, being inherently contradictory, cannot succeed except on the level of illusion, which is promptly refuted by the incipient perceptual acquisitions, it may be inferred that the neonate falls prey to a state that could be imagined, in adult terms, as terrible rage at being unable to make itself believe that the breast is in its possession, so that it hates the breast that it had begun to recognize as a good external presence. Where these information-processing circuits are exacerbated in their anti-mental direction, a pathological tendency can emerge, in which the attack on and destruction of the external breast are mixed with the experiences of sucking, which are transformed into angry, greedy sucking. Since this operation is unsuccessful, the fantasy of eliminating it is contemplated, and, given that a neonate can deal with this emotional situation only by splitting and projection, the external world is transformed into a monster that wants to eat him up. It may be postulated that a neonate experiences a relational state of this kind when, although hungry, he neither attaches himself to the breast nor sucks from the bottle, because these are perceived as extremely persecutory, bad and threatening presences.

However, the persecutory paranoid-schizoid world is constantly accompanied by the possibility of perceiving reality – that is, by the consolidation of perceptual phenomena. The upper part of Figure 6.1 can be continued as follows:

$$M+ \rightarrow B_{e_g} \rightarrow B_{i_g} \rightarrow P \rightarrow P+$$

In other words, out of the formation of an experience of a good external world (B_{e_g}), which leads to and is correlated with the possibility of a consequent good internal world (B_{i_g}), there arises the capacity to perceive that external world – that is to say, possibilities of perception (P) are outlined and consolidated. These arise simultaneously from the presence of experiences of good external objects (B_{e_g}) and from the intensity of good internal objects, which can be derived from their external counterparts through memory and the appropriate operations thereby implied. This corresponds to the reintrojection of the good object, or rather of the object that has been made good. *P* in turn further increases *M*, thus reinforcing a feedback circuit that leads to *P*+ – that is, to a progressive increase in perceptual capacity. This allows the mind to open itself to reality and become capable of perceiving it sufficiently accurately. It therefore gives rise to a mechanism antithetical to the paranoid-schizoid mechanism. If there were only the latter, the result would be psychotic hallucination, and there would be neither perception nor any possibility of real knowledge of the external world. The capacity for perception, although calling for biological apparatuses, originates not from these but from the first protomental vicissitudes – that is, from the operations which these apparatuses learn to use. These operations give rise to the progressive construction of perception, as analysed in Sections 3.2 and 4.5. With the gradual consolidation of the acquired perceptual capacities, these, as an effective instrument, come to constitute good parts, felt – this time stably – to belong fully to the subject himself. This represents the transition to the *eighth level*, involving the progressive consolidation of a self which, through its acquired capacity to perceive reality, becomes able to start effectively searching, in reality instead of in hallucination or imagination, for what it is now able to feel to be lacking. The increase in perceptual capacities and their utilization is reflected in a tendency toward knowledge, which characterizes the typical curiosity and search for new stimuli displayed by young babies after the sixth or seventh month of life, as well as the speed of the child's acquisitions from the end of the first and during the second year of life. The consolidation of this 'percipient self' can therefore be regarded as the transition from the protomental to the mental, and will render possible the unequivocally mental operations to be described in Section 7.3 (the ninth level), which characterize symbolopoietic development, the complex interlinking of fantasies and the reparative capacity of thought (see Chapters 9 and 10).

Let us now summarize the eight levels of protomental and, later, mental operations identified so far, specifying in each case the principal capacities progressively attained by the operational system:

1 triggering function of biological signifiers for the formation of the first anlage of the bad internal object
2 exorcism of the primary bad object by means of memory traces: autotomy
3 stabilization of mnemic-hallucinatory capacities and first anlage of a good internal object, which is hallucinated
4 organization of 'good = inside = self' / 'bad = outside = not self'
5 birth of perception and of the other-than-self

6 Ability to distinguish between memory, perception, imagination and hallucination
7 expulsion of the bad: persecution
8 consolidation of a percipient self and of perception of reality.

In addition to these eight levels, there is a ninth, which is fully mental (see Section 7.3), and characterized by the capacity to experience a bad self, with consequent possibilities of using the acquired capacity for reparation and hence, by means of the subject's own 'repaired' thought, for repairing the capacity to perceive, recognize and 'benevolently' know reality. This is a translation into cognitive terms of the Kleinian intuition of the depressive position.

The levels described above do not, of course, constitute developmental 'phases' or 'stages', but are intended merely as a logical framework to facilitate understanding of the operations learned by the mind system in order to become capable of any kind of knowledge. The classical clinical description in terms of affective states can be superimposed on these levels.

Chapter 7

Affects as cognitive operations

7.1 Envy and dependence

So far we have analysed the paranoid-schizoid position mainly in terms of logical operations that either permit or stand in the way of knowledge of external reality. A description of this kind often proves hard to understand. In particular, the following three difficulties arise.

The difficulty first is due to the naive and stubbornly maintained preconception that perception is an automatic process depending solely on the maturation of the neurosensory structures, whereby external objects are impressed as such on the mind except in cases of 'pathology' ('naive realism' – Katz 1951; Kanizsa 1980). Perception is in fact a complex construction (see Sections 3.2 and 4.5) involving learning operations; where it is found to diverge from reality, we should, before invoking intercurrent (such as affective or drive-related) causes, consider the course of the learning operations whereby the perceptual capacity was constructed.

Second, mental objects that have little or nothing to do with the objects of reality and their representations in the adult mind are not easy to conceptualize. The adult mind seems to use cognitive operations very unlike the primitive ones described, in so far as they give rise to quite precise mental symbols based on engrams that prove to be actual images of objects existing in reality, or at any rate images that correspond unequivocally to the objects and processes for which they stand – for the structures of logical-rational thought are based precisely on consistency and correspondence with the events of reality, and their validity performs an operational function in the external world.

The third difficulty concerns the use of the term 'internal object' to denote mental sets which – however unrealistically – signify external reality, given that psychoanalysts are accustomed to apply it only to internal reality. On closer examination, it is found that there can be no internal reality divorced from its comparison with external reality. Our approach describes protomental objects, which are therefore internal, but which function as overall signifiers of both external and internal reality, the two realities being confused and mixed up with each other in the signification. These protomental objects gradually come to be used, in the internal order, to differentiate what belongs to that order (operations performed there, which we equate with feelings and experiences, together with the separate engrams of imagination and memory); they also serve to distinguish the internal from the external (perceptual) order – the latter, like the former, being represented progressively more effectively by more specifically perceptual engrams. On the protomental level, therefore, any internal object contains within itself something 'taken' from the external world, which is used by the system for the purposes of comparison with that external world. There is no such thing as a wholly autochthonous internal object.[1]

Notwithstanding this progressive differentiation, the most primitive processes and engrams persist in the adult mind; psychoanalysis has described these as primitive or unconscious fantasies, anxieties and defences, conflicts, and deep feelings. The adult mind, considered solely in terms of its conscious aspect, appears not to use such engrams and primitive processes for cognitive operations – yet they remain active deep down. The conscious part of the mind does not register them as cognitive operations. The type of adult mind favoured by our culture tends to deny any cognitive function to the primitive processes; at most, a limited value as 'feelings' is attached to them. This makes for considerable difficulty in understanding the description of the most primitive processes, especially if they are described not simply as 'impulses', 'fantasies' or 'feelings', but in terms of mental operations which, whether correct or incorrect, are stated to be 'cognitive'. Hence the difficulty in, or resistances to, full comprehension of our description. Again, the protomental processes have to do with non-existent, ineffable objects – that is, ones that cannot be described in verbal language – and with illogical operations and processes. Since we must nevertheless have recourse to words in attempting to describe them, we are bound to use circumlocutions, metaphors or terms that are in one way or another inappropriate for their representation; a discussion of protomental processes cannot but be difficult to understand.

The problem is still further complicated when we attempt to move on from the description of engrams and processes concerning contact with the external world (perception being the most elementary example) to that of the more internal experiences – perceptions, so to speak, of one's own processes and thoughts: the engrams used by the mind system to 'think its own thoughts'. The internal objects as classically described by psychoanalysis surely exemplify how the subject can have experience of parts of himself; here

1 Unless it is again assumed to arise out of Freud's 'witch metapsychology' (Fabozzi and Ortu 1996).

too, then, we are concerned with cognitive, or protocognitive, operations, which could perhaps be defined as unconscious self-cognitive operations.

It may thus be helpful to repeat our earlier description of the first internal objects that was given in terms of mental operations, but this time from the standpoint of feelings. The latter approach, or 'vertex', is, moreover, the one first used by psychoanalysis in considering such phenomena. By correlating this second way of describing primitive mental events with the first, which has thrown a different light on them, we shall be able to show how the neo-Kleinian theoretical descriptions of the first object relations and feelings that inform the psyche of small children agree and overlap with, and indeed constitute a particular description of, what we have described here as the first mental 'operations'.

Let us therefore now analyse the paranoid-schizoid position from a different 'vertex', which, while more familiar to psychoanalysts, is liable to be misunderstood by other psychologists with their somewhat 'adultistic' vision. This vertex is that of an infant's first object relations and feelings. This description is given *after* the previous one precisely because anyone approaching it with a merely theoretical psychoanalytic background (that is, without the experience of a personal psychoanalysis) is bound to understand it on a summary, or 'adultistic', level, in which Klein's technical terms are unconsciously understood as referring to conscious adult feelings. The terms proposed by Klein refer not to feelings that can be experienced by an adult, even if they use the same vocabulary, but to ineffable experiences such as, precisely, 'fantasies' or, from a different point of view, the protomental 'operations' described in the previous chapters.

Envy, defined as the wish to destroy something that appears good and desirable and is seen as unattainable (Segal 1964b), has been regarded as homologous to feelings of hate and as the primary subjective expression of the destructive drive or 'death instinct' (Freud 1920g, 1923b, 1933a [1932]; Heimann 1952; Money-Kyrle 1955, 1961). Klein (1957) pointed out that envy is an attitude that arises very early on in human life; in her view, a child begins to envy as soon as he comes to perceive the existence of something external to and different from himself. A neonate thus envies the first object that presents itself as something good external to himself – that is, he envies the 'breast'. The recognition that the breast is something external on which his own well-being depends gives rise to envy, rage and hate, and, given the impossibility of totally possessing it, also to the wish to destroy it. Now Klein misused the term and concept of a 'wish' in accordance with an adultistic image: the hypothesis that an infant primarily envies what he perceives as different from self presupposes a wish for possession. The assumption or consideration of a wish to explain an envy referable to the first months of life is tantamount to attributing to the infant mind at this stage something mental that the child attains only later, when he comes to recognize a need as an unpleasurable state of his own and conceives the possibility that an external object might afford relief, and might perhaps also be reached in some way by him. The recognition of all this would mean that the baby has already moved on, as Winnicott (1965a) points out, from a situation of 'double dependence' – in which his survival depends wholly on the mother, although he does not recognize this, living as he does in an illusory state of omnipotence – to one of 'dependence', in which the link is recognized and experienced in its aspects of satisfaction, vulnerability and suffering. Winnicott calls the primitive infantile dependence 'double' because its non-recognition doubles its significance: in so far as a neonate does not recognize his dependence on others, he will be unable to draw their attention to any of his needs and will depend not only on their intervention, but also on their capacity to foresee the need he does not communicate.[2]

Envy is, as it were, the opposite of the recognition of reality – that is, of the recognition of dependence. It tends to preserve the child's illusion of his own omnipotence, so that, as soon as he notices that the breast is not a part of himself, but of external reality, and that it is something good and to be wished for, but is not completely and permanently attainable, he hates it; in these terms, the infant may be said to hate and attack the 'breast'. However, this is not a primary process: it can occur only at a later stage. Whereas projective processes concerning feelings such as hate, envy, aggression and their consequent persecutory nature can be described, on the primitive level processes of projection and perception are more readily describable not with reference to complex mental configurations such as the feelings mentioned above, but to more elementary mental 'forms', or formless engrams, which are just as effective in producing contradiction and confusion – the confusion that could later be called rage – when attributed to the external world as when laid at the door of some kind of 'self'. We may speak of a persecutory breast created by projection – a breast that is greedy, envious and devouring, and is thus fantasized in oral terms – but this is a later process, which takes place as and when the infant learns to use memory in its realistic meaning, to approach external reality and to recognize objects. Although the persecutory object can be called a 'breast', this usage may be inappropriate; furthermore, it is not just one object, but a number of configurations, even if the common denominator of this multiplicity may be a 'form' derived from the vicissitudes

2 'The great change that is noticed in the first year of life is in the direction of independence. Independence is something that is achieved out of dependence, but it is necessary to add that dependence is achieved out of what might be called double dependence. In the earliest stage there is no vestige of an awareness of this dependence, and for this reason the dependence is absolute. Gradually dependence becomes to some extent known to the infant who, in consequence, acquires the capacity for letting the environment know when attention is needed. Clinically there is found a very gradual progress towards independence, with dependence and even double dependence always reappearing' (Winnicott 1965a: 4f.).

of oral experience. At any rate, when these persecutors are formed, the infant fears that they may somehow enter into him. The situation escalates: the worse any persecutory external 'breast' becomes by virtue of projective mechanisms, the more feverishly further projection will be attempted. This could be called fear or threat, but these terms are perhaps too adultistic for what happens in the infant mind. From the point of view of protomental operations, the situation is paradoxical, in that the inappropriate mental operation, which put the baby into this state of persecution in the first place, is used in a futile attempt to eliminate it, thus creating a persecutory breast raised to the *n*th power (see Section 6.4). All these mental dynamics, taken together, in my opinion, offer a better description of the envious attack. This circuit can be interrupted only to the extent that external reality does not become worse and worse, but can be experienced in its good aspects too, thus allowing the inevitable pain to be tolerated and preventing the projective mechanisms from being short-circuited.[3] In other words, the problem is now that of potentiating a good internal object by reinforcing the perceptual memory (*PM*), which means that the effects of the external reality, or environment – that is, the mother and her specific function of reverie (see below) – must be taken into account. The more this function reinforces the child's experience by the presentation of sensory stimuli that have already begun to be recognized as good sets and objects, the more the set of memories – that is, aggregates of afferences with psychic significance – will be enriched; and the more significant and non-equivocal these memories are, the more they can contribute to reinforcing the formation of an internal object and a memory that can be used for effective operations that can offset the persecutory dynamic.

7.2 The good breast and the capacity for memory

In the situation we are attempting to describe, the infant begins to have a kind of experience of something good that belongs to himself and that can correspond to something external, although not being identical to it: it is everything that has been called the 'good internal breast' – that is, the good internal object, or rather the set of good internal objects, formed as engrams of memories of happy experiences in relation to the child's caregivers. This store of 'memories' not only serves to increase his capacity to recognize the external world as opposed to merely hallucinatory events, but also enables him to wait, without despairing, for the satisfaction of a need and to embark on a process of coordination – for example, to modulate his crying – in order

3 Where the external reality is particularly and objectively painful, as in the case of neonatal traumas and diseases, the construction of the mental apparatus is at risk; hence the caution that must be exercised in all neonatology and paediatric wards.

to obtain what he can begin to wish for. If this set of 'memories' has become sufficiently stabilized in the baby's mind, he will also be able to recognize the set of external circumstances that immediately precede feeding. A 2- or 3-month-old baby is capable of communicating with his mother as she performs the preparatory actions to feeding him, and of recognizing that the memory of previous experiences is about to be translated into reality; he will then be able to tolerate hunger without turning it into something bad that must be projected. The smile that begins to appear on a baby's face in the second month is, precisely, a signal that he has begun to recognize something external to himself and is using the memory to enter as far as possible into communication with this external object.

The smile usually directed by a 2-month-old baby (Spitz [1965] placed this event in the third month owing to his study population of relatively uncared-for, institutionalized children), not only to his caregiver but also to stylized representations of the human face, has led some to consider that the smile reaction may be innate, like a newly hatched gull's pecking at a red spot in a white field. Yet, the eyes–nose–mouth gestalt, although consistently present from birth, is not recognized before the second month: the baby was therefore previously unable to perceive any visual form.[4] The smile reaction thus proves to stem from experience and from its organization on the basis of a minimum of mental structuring. All subsequent finer discriminations will be later achievements: at 5 or 6 months, the infant will be capable of distinguishing different faces, and toward the seventh month will smile at family members and cry when confronted with a stranger (Spitz's 'eighth-month anxiety').

The phenomenon of stranger anxiety, which, moreover, is not universal, can be conceptualized in our terms as the failure to recognize and refind the good object. If an infant sees a human face and does not recognize it because it cannot be accommodated within the order of engrams belonging to the good external breast, he may well project his persecutory internal fantasies on to the experience. The more disturbed the child, the more clamorous this reaction will be, whereas an infant with a sufficiently well established internal good breast can tolerate the experience of the stranger without attaching his persecutory projections to it.

A child who has learned to recognize the good maternal presence in the external world can tolerate waiting and preserve hope – that is, the internal good breast. However, this acquisition has quantitative limits: if hunger increases or some external condition accentuates the state of discomfort, or if waiting is prolonged, the infant will no longer be able to endure the pain by

4 Peripheral maturation (of the retina and visual pathways) is well under way, but central maturation requires learning operations; cf. the concept of 'maturation' and its popular misconceptions.

maintaining the engram of imminent relief, so that the projective circuit will be reactivated. Hence, the relationship between realistic and projective mechanisms of reality recognition can be seen as a dialectical process that is not exhausted in a single transaction, as the child's mind oscillates continuously between the states induced by the two mechanisms. In this way, two contrasting levels of mental functioning are established, one oriented toward the recognition of reality and the acquisition of good experiences, and the other toward their denial. On the one hand, the child has the sensation that the 'breast' is something external, which he can recognize and hence obtain, while, on the other hand, he experiences the 'breast' as something of his own, which he can summon up at will to do his bidding; together with the recognition of external reality, a feeling of hallucinatory omnipotence persists, and can readily flow into persecutory circuits.

The first mode of functioning involves an increase in perceptual capacities and in the stock of memories, in the sense that it favours the preservation of memory traces less and less imbued with persecutory fantasies, hallucinations and apperceptive distortions, and increasingly appropriate for representing external objects as they are, or at least in a manner more closely resembling their actuality. In this way, memories become representations similar to those of adults – that is, ones that adequately reproduce the objective characteristics of external objects. The mind thus acquires the possibility of a more appropriate match between its own internal objects on the one hand and external objects on the other hand. All this increase in memory and perceptual capacity allows the mind to begin to function in terms of reality-matching operations – that is, operations of actual knowledge. Since knowledge allows the child to endure states of uncertainty (anxieties) and to acquire new and increasing good experiences, knowledge itself takes the form of a good object, a good internal breast, so that thought processes – that is, all 'appropriate mental operations' – constitute a good experience for the mind, and therefore continuously enrich the 'good internal breast'.

The constitution of mental processes as good objects – and of certain processes in particular, namely, those effective for interacting with reality – proves to be a fundamental factor in a mental anabolism, in the sense of a tendency toward ever richer and more effective symbolopoiesis, as opposed to the catabolism inherent in autotomy, in the avoidance of unpleasure and, in general, in the first levels of protomental operations.

Alongside this new order of mental functioning that is in the course of becoming established, the previous level continues to operate, and between the two there arises a perhaps infinite series of intermediate processes, in which cognitive processes proper, constituted by correct mental operations, alternate with, or are mixed with and superimposed on, others that are cognitive only in a much looser sense: attempts at cognition which, however, do not culminate in reality-matching operations, but consist of fantasy processes like the primitive ones described earlier as 'incorrect operations'. In parallel with internal objects that match their external counterparts – that is, representations more or less similar in conformation to the relevant external objects – the mind preserves all the preceding internal objects, which are of the order of fantasies and accordingly have little or no correspondence with anything actually existing in reality.

The mind is thus seen to possess aspects of two different worlds. This first contraposition of levels may be postulated as being the primitive matrix of the later structural differentiation into one level of more strictly cognitive operations, which will subsequently become partly conscious, and a second level of irrational operations that will give rise to the world of feelings, emotions and passions and, in adults, will remain largely outside the sphere of consciousness. In this way, adult thought will ultimately manifest its twofold aspect: on the one hand, the conscious rational epiphenomenon and, on the other hand, all the unconscious work that underlies every conscious thought. Bion (1962, 1963, 1965) describes the relationship between the two levels in detail.

Conflict too, as described by psychoanalysis,[5] can be conceptualized in terms of a dual level of mental operations, the first level comprising realistically oriented mental operations and the second those based on unpleasure avoidance – hallucination, projection, evacuation, persecution and, later, wishes and anxieties. Hence, the initial conflictual situation can be postulated to arise at the point when the infant, although he has begun to perform realistically oriented mental operations, continues to resort to hallucinatory modes of thought.[6]

5 The notion of conflict in psychoanalysis arose in the framework of the energy-and-drive theory (forces in conflict), and remains stubbornly attached to this conception, so that some find it difficult to approach it differently. Peterfreund's attempt to eliminate the terminology of energy from psychoanalytic language (Fabozzi and Ortu 1996) is relevant here.

6 This is also the locus of the problem of conscious–unconscious differentiation or, put more appropriately in our view, of why certain mental processes can assume the character of consciousness. Without digressing excessively, we may say that correct, realistic operations are more likely to be able to become conscious because they are unequivocal, suitable for appropriately representing what exists and is happening, and are therefore verbalizable; incorrect, or inappropriate, operations, on the contrary, are confused and confusing, preverbal and ineffable. The two kinds of operations take place in parallel even though they are mutually contradictory. Consciousness can arise only when the realistic operations attain a certain level; it perhaps consists of operations that have attained a certain degree and form of unequivocality and of correspondence with both external and internal reality. This is the sequential processing described by the cognitivists, as opposed to the kind of parallel processing that intrinsically *cannot* be conscious (Imbasciati 2004b: Chapter 9). This manner of describing, or rather, here, of explaining, the unconscious, may puzzle some analysts, who may feel that it detracts from the determining value and force of the unconscious. The difficulty lies precisely in the concept of 'force' – that is, the fact that the concept of the unconscious remains bound up with Freud's energy-and-drive theory. What actually calls for explanation is not the unconscious but why the subjective character of consciousness arises at all in certain cases (Imbasciati 1989b). I have argued elsewhere (Imbasciati 2001, 2002a, b)

The differentiation of the two modes of functioning (and, in general, of the eight levels identified earlier) should thus be understood in the sense of a dynamic interaction, whereby the irrational part becomes a source of elaboration for the logical and rational part; moreover, this dialectic between more strictly cognitive processes and affective processes will persist in the adult mind. In the course of development, therefore, we observe a trend toward greater organization and extension of the realistic operation of the mind, even if the possibility of reverting to more primitive operations persists.

7.3 The 'bad self' and experiences of a thought of one's own

With the establishment of the experience described here as the good external breast (B_{e_g}), the child's mind seeks to acquire it for itself, by all the means and in all the configurations possessed by its fantasy ramifications, which are of bodily and sensory origin. The infant can therefore be said to want to 'swallow' all the 'breast', or 'breasts', which he remembers or imagines to exist. The mind is capable of discerning the discrepancy between this incipient 'wish' and reality: it can observe that the good external object disappears, so that the state of need is prolonged (for instance, the bottle may be empty, or the nipple may be withdrawn from the mouth). This may give rise to a kind of fear that the child's own state of need has caused the breast to disappear or that the operations the mind was attempting to organize so as to make a realistic approach to the external object may instead have destroyed it or caused it to disappear, or 'consumed' it. The primal hallucinatory illusion of being able to create good things is now accompanied by a new experience – namely, that of being able to consume or somehow to destroy them.

The infant has acquired an initial capacity to perform a reality test on the external world: an external object capable of catalysing this test may, but may also not, exist as a presence in reality. The test now becomes one applied to the internal processes which the mind is organizing and beginning to experience as its own and as generated by itself, unlike the incipient representations of objects that are beginning to be experienced as external and as not belonging to the self. 'I notice that the breast is no longer there' becomes 'I notice' something that has happened in the mind – a 'thought' – and this constitutes the formation of the experience of the self as made up of mental processes. When the reality test produces a distressing result – 'the breast is no longer there' – the test of a 'thought' might be performed: 'I thought of a breast, but

that the question that underlay the birth of psychoanalysis – 'why the unconscious?' – is wrongly stated; we should instead ask, 'why consciousness?' and change our psychoanalytic theories accordingly.

it is not there; I only thought of it.' If the distress cannot be endured, there is the possibility of experiencing as distressing the actual mental process that leads to the recognition of reality. In this case, instead of noting a thought, the infant may regress to the use of hallucination. However, a further, new operation is also possible, definable as a kind of doubt that the infant's own mental processes, whose goodness and validity had been verified by the observation of the present 'external breast', might instead have caused it to disappear. This could be described as the initial embryo of a sense of guilt.

On the other hand, alongside the processes directed toward the recognition of reality, the persecutory processes persist; these are tantamount to destruction of the good objects constituted by the 'appropriate' mental operations themselves (perception, memory, etc.). This destruction, carried out with a view to avoiding suffering, has instead itself become a source of suffering and tends to be evacuated by the self, but is then liable to come back inside it. The process of evacuation and projection, strictly speaking, is never perfect; it therefore leaves behind a kind of experience of these bad parts as actually constituting the self, despite the wish to expel them. In this way, autotomy basically comes to be experienced as the highly destructive and terrible event that it is in reality, constituting as it does the death, piece by piece, of the nascent mental apparatus. This situation, together with the testing and fearful doubting of appropriate mental operations as described above, constitutes the nucleus of an initial experience of self as a possibility of generating something unpleasurable, owing to insufficiency or 'error'. There forms the nucleus of a primitive sense of guilt – a self that can begin to be experienced as 'bad'. It is felt to be bad because capable of attacking and destroying the subject's own nascent 'good' structures and, with them, the possibility of actually reaching an 'object' of some kind.

Here we may identify a *ninth level* of protomental operations, or rather a further, more properly mental, step beyond the previous eight levels, which were wholly protomental (see Section 6.5). We have already described the acquisition of the mental capacities, which are experienced as a good object constituting the self (Section 7.2), with consequent symbolopoietic anabolism, and the observation of the subject's own thought with respect to reality (see above), with the consequent first glimmer of the intuition of a self constituted by 'thoughts'. Now, however, there is also the experience that thought may be 'bad' and give rise to guilt. This appears to be a turning point in the structuring of the self – a good, effective, percipient, knowing and thinking self, but also a self that can be 'bad', and capable of attacking thought.

Besides the experiences described above, there may be others, which tell the child that attainment of the external 'good' may be accompanied by pain. Among the most common is stomach-ache associated with the ingestion of food. Such an experience conveys the message that there is something bad inside; if the child no longer believes, in accordance with the primitive schizoid mode, that an evil spirit has entered into him, and if he no longer makes the projective mistake that the breast given to him has been bad, then he is, so to speak, compelled to experience the 'bad' as originating from himself. The problem now arises, for a mind that has grown stronger by experiencing a good nucleus of the self (a good internal breast, equated with knowledge-related processes), of being able to feel itself destroyed; that is, to feel that it can be destroyed by 'bad' operations, by a bad experience, which may be internal – that is, one experienced as a bad internal breast, this time felt to be its own. A bad internal breast of this kind can be denoted by the overlined symbol ($\overline{B}_{i_b}$) to distinguish it from its predecessors (B_{i_b}), which tended to be eliminated and expelled from the self; this one, however, is the first bad internal breast felt truly to belong to or be constitutive of the self.

To sum up, this ninth operational level can be expressed by the following notation:

$$\left(B_{e_g} + B_{i_g} \rightarrow \overline{B}_{i_b}\right)$$

This expression indicates that, in so far as the mind becomes able to have good experiences in relation to good objects existing in external reality, and hence to experience its own nascent mental structures based on effective operations as good, there also arises a first experience of 'badness', felt painfully to belong to the same self in which there may be good objects. The mind must then confront the problem of keeping good and bad objects together inside itself, distinguishing them from each other and no longer processing them in accordance with the primitive modalities of 'everything good = self' and 'everything bad = outside-the-self'.

7.4 Splitting and idealization

We have described the process of construction of the persecutory breast, which tends to return inside the child and to increase his fear and suffering, thereby establishing the vicious circle of persecutory projections. One of the ways in which the child tries to defend himself is a mental operation which, while substantially false, nevertheless leads to temporary alleviation of the pain – namely, the splitting and idealization of objects.

The establishment of a good internal breast effectively lays the foundations of mental life and is correlated with the constitution of the – first external, and later internal – bad breast. Both organizations, good and bad, stem, as stated, from the same experiences: one and the same experience, for instance, that of sucking, will be 'good' in so far as the baby's mind recognizes the experience derived from the 'breast' as good, and 'bad' where the same sucking-related afferences 'remind' it of hunger, frustration, unsuccessful exorcistic hallucination, chagrin, rage, hate and the envious attack, which become coagulated, by paranoid-schizoid mechanisms, with the formation of everything 'bad'.

If the 'good' continues to be experienced as originating from an external object (B_{e_g}) by virtue of consolidated processes of perceptual recognition, and the good belonging to the self is experienced, not in the concrete object, which remains external, but in the mental processes which make it possible to recognize and seek it out and in those whereby it is used with the result of gratification, there then also forms within the self, and is felt to belong to the self, the experience of something bad (or 'incorrect'), which the child is unable to set against the 'good' that was in the process of formation inside. The child presumably now lives in fear that this bad internal breast might overcome, swallow up and submerge the good one, or become confused with it. He will then attempt operations directed toward distinguishing the 'good' and 'bad' inside himself, or rather toward keeping them separate, one remote from the other. The adult mind, or at any rate the more developed mind, knows that this internal good breast is the representation (perception or thought) of the object that is outside, and that what is bad is our reaction of anger toward the object because it has sometimes disappointed us by not being present, not arriving instantly, or not being available for our possession at will – in other words, it knows that the good representation of a good object is one thing and that a painful ('bad') feeling toward the same good object is another. It is one thing to know that one can 'know' the good, but another to feel that this possibility of knowledge can be 'reversed', in an anti-mental dimension destructive of one's own thought and of all knowledge. The distinction between good and bad – both internal – depends on the capacity to distinguish the mental representation of an experience from our emotional reaction to this representation. However, this distinction cannot yet be made by a very young baby's mind, which is unable to distinguish the mental content from the mental function that accompanies it; for the neonatal mind, both are 'objects'.

Both, as it were, preserve the 'substantiality' of the external object from which they are derived, and both are homologous with 'contents' or representations. If a mental function such as 'hating' the breast (the fantasy of tearing it apart, biting it, annihilating it, etc.) is active in the infant, this function is itself an object fused with the breast object. The bad breast is thus a breast that is bad because the child wanted to attack it, swallow it, bite it and tear it to pieces, and it is likewise a breast that tears to pieces, bites and gobbles up his mind; there is no distinction between active and passive, just as there is none between an action and its object. If the baby no longer attempts to expel the bad internal breast, he finds himself compelled to keep it as distinct and separate as possible from the good internal breast, in order to preserve (at least in part) the good experience that promotes well-being, security, hope and knowledge from contamination. In this way, an experience that refers to a single real event, feeding, is artificially split into two experiences. This operation is not realistically oriented, but has the sole purpose of preserving the existence and vitality of the good internal object. This is achieved at the cost, at first, of failing to recognize the bad object as belonging to the self and, in this second phase, of keeping separate within the self, as if they were two profoundly different entities stemming from utterly diverse external experiences, two 'mental objects' that are conceived as different although they in fact relate to the same external object. However, if B_{i_b} is no longer expelled, it cannot be equated with B_{i_g}, and the two cannot be experienced as the same entity toward which the mind has developed two different attitudes. For this reason, the possibility of this equalization instead leads ultimately to the separation of the two entities, in accordance with the following expression:

$$B_{i_g} = B_{i_b} \rightarrow B_{i_g}\Big)\Big(B_{i_b}$$

The external object is therefore also split in two; that is to say, the same external 'breast' object – in other words, the same experiences of sucking – is split as if it referred to two external objects, one bad and one good:

$$B_{e_g}\Big)\Big(B_{e_b}$$

For the more developed mind, such an operation – especially its final part – seems absurd. The object constituted by the nipple, bottle or whatever, is always the same; in other words, it is always perceived as such, whether we are well disposed to or angry with it. For the neonatal mind, however, there are two objects, one good, which feeds, is recognized, loves and is loved, and the other bad, which starves, devours, is hated and hates, and can persecute. The infant cannot distinguish the content and representations of an object from the function of his mind and from the relationship linking them; the two orders of experience are two different objects, which are kept separate and split. Only when the mind becomes able to unify the two objects into a single one will a totally new differentiation come into being. This will constitute a significant achievement for the child – namely, the capacity to distinguish between his own mental representations of external objects and his own feelings toward them, which is tantamount to recognizing hate, envy, aggression and all the 'bad' fantasies as his own.

This transition, which will constitute the very cornerstone of the cognitive capacities, will be described in detail later, in our discussion of the depressive position. Here too, the transition is from protomental to more properly mental operations. Previously, the child kept good and bad objects separate, both inside himself and in the external world; but now, the bad objects are felt to be wholly bad, whereas the good ones can be experienced as sublimely and utterly good – in other words, they are 'idealized'. In this sense, the mental operation of splitting and idealization is one that it not realistic but is directed toward the preservation of good internal objects that permit, and will permit, processes of contact with reality. Persecution would otherwise gain the upper hand and the total absence of good internal objects would not allow any contact with reality or any constitution of the self, which would fall prey to continuous autotomy.

A further aim of splitting and idealization is to postpone the 'vision', which would be intolerable to the mind and therefore repudiated, that, if the bad internal object is an attack on the good external object, this bad internal object really is trying to harm the good breast – the external one and, with it, also the internal one derived from it – and could cause it to disappear. That is to say, 'incorrect' operations (a kind of negative proto-thought) really can destroy the nascent good cognitive structures. Thought itself, in its first, purest – and realistic – primary expression, namely, perception, could indeed be destroyed, if contaminated by the bad internal breast – that is, by the mental operations underlying what we later infer to be destructive feelings. In other words, like a computer virus, the incorrect operations erase the mental structures that have already been constructed or are in the process of construction. That is what happens, for example, in psychotics, in whom the perceptual processes are conspicuously compromised.

Splitting and idealization thus represent a stage in the normal development of the mind: if the child is abandoning the hallucinatory processes whereby everything good is self-created and the bad is outside, the fact of containing within himself both good and bad entails incipient experiences of intense anxiety and, in particular, the fear that the bad now experienced within the self might contaminate the good, thereby destroying the subject's own 'good' mental structures. To cope with this anxiety – or proto-guilt – the infant preserves the good things by paying the initial price of excessive separation from the bad ones; the split-off parts cannot be recognized, and therefore remain active as potentially destructive entities whose malignity is unseen, so that they cannot be 'repaired'. Splitting and idealization thus constitute an intermediate stage between the hallucinatory and exorcistic processes, on the one hand, and, on the other hand, those of putting together the internal good and bad, which will be described as reparative (Chapter 10).

Conceptualizing these processes in terms of anxiety, we may discern a transition from a proto-anxiety, due to the appearance of the 'bad monster' (see Section 5.1) and the consequent exorcistic operation, to paranoid anxiety and unsuccessful persecutory defences, and thence to the confrontation of a proto-guilt – another intolerable anxiety – by processes of splitting and idealization. The anxiety will subsequently be metabolized as guilt when the good and the bad are set against each other in the depressive position.

Hence, the processes of splitting and idealization are deployed for the transitional purpose of enabling the mind to attain the more elaborate and creative cognitive operations – namely, reparation – that constitute the decisive shift toward symbolopoiesis. Alongside all the more developed operations arising out of this, however, it is important to remember that the initial modalities remain active in the functioning of the adult mind; they can often be discerned at the root of what we call attitudes, feelings or, sometimes, character traits. For example, splitting and idealization may remain active in mental situations in which the objects, persons, things, ideas and social forms of external reality, as well as the subject's own thoughts and feelings, are experienced as either wholly good or wholly bad – that is, they are either idealized or execrated. In other words, objects, both external and internal, which in the majority of cases have both good and bad aspects, are split into their two components, one or other of them being denied, projected or displaced on to other objects. This happens because, at a deep internal level, it is feared that the badness bound up with an object might overwhelm and contaminate it, destroying any goodness it possesses, and that rage and envy might attack the very mental structures, which are essentially good objects to be preserved. In this way, of course, rage and envy are concealed.

Some people customarily exhibit processes of splitting and idealization, while others who do not normally do so may resort to them at certain times. For instance, an adolescent, owing to his intensified sense of guilt, may readily espouse a Manichaean-type attitude in which he draws a sharp line of demarcation between good and bad, idealizing certain objects, persons and ideologies and execrating others, without realistically comparing things that, in reality, are not as apodictically antithetical as he sees them (Erikson 1968; Josselyn 1971; Grinberg and Grinberg 1975). In some situations, even the mature mind may be driven by passions and feelings to make a drastic separation between good and bad and to idealize what is felt to be positive. The tendency toward idealization may be exhibited in all situations of intense personal involvement – that is, whenever all the deep mental and hence protomental processes are activated – as in the state of being in love or in certain forms of ideological commitment. The aim, at a time of prevalence of a particular mental economy, is to preserve the good internal objects from the subject's own destructive attacks. Hence, idealization is often indicative of an intensification of destructive fantasies and of the dialectic between paranoid-schizoid processes and the reparative processes essential to offset them. The subject would otherwise run the risk of 'going mad'. Recourse to splitting and idealization in adult life is therefore a sign that, to confront an internal emergency, the mind is once again compelled to keep good and bad separate, with a view to preparing further, new reparative processes to reinforce, or re-create, the internal good.

Chapter 8

Corporeity and modes of thought

8.1 Engrams: sensoriality and memory of functions

The first engrams (protoengrams) are formed as units for the reading, and hence also the recognition, of configurations of sensory afferences.[1] This is already the case in the fetus (Della Vedova, Manfredi and Imbasciati 1996; Imbasciati 1997; Manfredi and Imbasciati 1997; Manfredi, Tommasoni and Imbasciati 1999) for hearing, for the chemical senses and partly also for the sense of sight (light intensity). The same is true of the tactile-proprioceptive senses, which entail integration with motor efferences; in other words, the engrams are also based partly on the brain's commands to muscle groups. The successive motor acquisitions, already in the fetus and, to an even greater extent, in the neonate, show that the motor programs (outputs), in order to be recorded (memorized) in their progressively increasing sophistication and complexity, must be integrated, in the course of processing, with the progressively growing capacity to process inputs. These inputs are continuously 'calibrated' against the afferences relating to the progressive operations whereby perception is constructed. Hence, the protomental operations use not only successive engrams of sensory origin, but also 'internal' engrams, made up of the psychomotricity that is in the process of construction, as well as all the internal engrams that represent the first mental 'products', such as affects, as described in the previous chapter.

The process of learning to perform progressively more complex (protomental and mental) operations, which can be described, for the sake of simplicity, principally in terms of the organization of sensory inputs into engrams endowed with meaning, thus takes on a broader complexion, involving the progressive, complex construction – and hence also memorization – of

1 This does not mean that the recognition of configurations corresponds to a perception proper – that is, one that matches the reality of the configurations as an adult would perceive them. In other words, it must be borne in mind that the assembly by which they are read may not correspond at all to a perception in the strict sense (see Section 6.5).

new and complex engrams that correspond to the progressively increasing operational 'capacities'. The engrams of sensory origin are therefore integrated with successive 'internal engrams' in the complex progressive creation of new symbols – the process of symbolopoiesis. Among the most important of the various internal engrams are sensorimotor integration and the recording of programs of efferences: afferences and efferences are used to construct the first internal products, the first thoughts, and these – whether memories, imagined entities or increasingly complex connections (actually the increasingly sophisticated capacities of the mind in the process of construction) – in turn contribute to the reading and subsequent organization of every input received. This process underlies the progressive capacity for symbolization that characterizes the more developed mind, as well as its operational capacity.

In parallel with the above *description* in psychological terms, we can offer a psychophysiological *explanation* (see Section 2.1), whereby all the possible engrams, or signifiers, are related to corresponding 'traces' (see Sections 2.4, 3.5 and 4.2). These must be seen as including, first and foremost, the traces underlying the processes themselves – that is, the progressive acquisitions of capacities to process all the (mainly internal) 'information' created and circulating within the system. The engrams should therefore be conceived not as images of compositions of afferences, or of compositions of afferences-cum-efferences, but as traces of functions – that is, memory traces whereby each successive function acquired for working with the available engrams, such as the various levels of protomental operations, must be acquired or, put differently, memorized. Symbolopoiesis concerns not only mental 'contents', however absurd or inconsistent with reality they may be, but also the production and recording in memory of progressive operational capacities; hence, this memory is one that stores functions.[2] The more complex the sets of multiple combinations of afferences and efferences – resulting from the exercise of bodily parts and organs – used to construct them, the more complex these functions will be.

In parallel with this progressive construction of symbolization processes, infant observation reveals a progressive and often consecutive use of various bodily zones, with respect both to their specific sensory apparatuses and to their characteristic motricity. For many years, psychoanalysts have invoked the concept of 'organ modes' (Erikson 1950); indeed, Freud himself, in a different context, described the 'phases' of development in terms of orality, anality and genitality. The part of the body that is exercised determines the function by virtue of its anatomy, and hence also the functional program that is learned, as well as the specificity of the operations and associated engrams that are constructed. The specificity of the operations is then extended beyond

2 This memory is therefore procedural in the broadest sense of the word.

the particular organ whose exercise gave rise to it. In this connection, it may be wondered what determines the observed succession – for instance, why motricity comes after the organization and acquisition of certain operational capacities related to the senses (mainly those of hearing and sight). Rather than adducing the biologically determined maturation of mental psychomotor apparatuses, it seems reasonable to assume that motricity can begin to be elaborated only when the mind, initially working on its sensory inputs, has acquired certain levels of operations (first and foremost, perception) allowing the use of proprioception, together with other sensory inputs, to *construct* the first programs of efferences that constitute the first psychomotricity. These will therefore permit the exercise of certain bodily zones, which will in turn contribute to guiding the construction of subsequent programs.

Development consists in the progressive construction of capacities for symbolization – both of what is received from outside and of what is produced internally – and may be carried out or integrated in some way with the received input. In this way, the organ mode becomes a thought mode, or, at least, a proto-thought mode.

So far we have described mental operations in terms of aggregates derived from feeding-related experiences, in which afferences from various senses are mixed together and confused (see Sections 3.2 and 4.4), so that the mental objects are very different from memories in an adult. In adults, both the operation of remembering and its content consist of images which, apart from being more precise, are made up of traces of diverse sensory origin (mainly visual); however, these are organized in well-defined sets and connected with highly particular auditory memories – namely, words, with which we can describe our experience. It is therefore worthwhile to examine the 'stuff' of which the first mental processes are made – that is, to consider not only the logical terms of mental operations, but also the *elements* with which mental operations work. These elements are the traces of the sensory afferences that gradually come to be organized in memorizable sets suitable for use by the operations currently in hand. More generally, the various sensory apparatuses and bodily organs supply the material with which mental operations will be performed, and, depending on the nature of these afferences and on the conformation of the organs that generate them, the manner in which the mind is able to process – or 'metabolize' – them may be influenced in consequence. Hence, the physical organs and the specific form of their inputs influence the models that the mind can construct. The succession in the use of the various traces from different parts of the body will depend on the extent to which their specificity is integrated with the level of operations so far attained by the mind and, conversely, also on the extent to which this level can in turn be influenced by the use of the various afferences (Bernfeld 1925; Erikson 1950).

We can therefore not only examine the vicissitudes of feeding and – as we have done – describe the development of the traces originating from the relevant bodily zones (and thus refer to the internal objects in 'breast' terms), but also apply this type of description to the use of other bodily zones and sets of afferences, so as to see how they cause the level of possible mental operations to develop. In other words, we can also describe the processes of symbolization subsequent to those of oral origin. We have already mentioned the processes of fetal symbolization (see Section 4.1) and noted that, if these were sufficiently well known, we should have to revise our entire conception of the subsequent construction of oral symbolization, whose origins we have, somewhat arbitrarily, attributed to the vicissitudes of orality. At any rate, even if this project is deferred to some future date when more may be known about the construction of the fetal mind, we can already, with the knowledge at our disposal, describe post-oral forms of symbolization on the basis of the stamp impressed on them (see Sections 2.5 and 4.1) by the preceding operational constructions, which stem, precisely, from oral experience.

I have sketched out such a description elsewhere (Imbasciati 1983b).[3] Some of its salient points are recapitulated in this book, in order to show how the operations with whose progressively higher levels we are here especially concerned make use of increasingly interlinked material – internal objects – for the successive commencement of functioning of other bodily zones, with the use of the relevant experience. Learning from experience proves to be a matter of learning from the body.

Let us now briefly consider the auditory experience, the visual experience, the experience of prehension and the experiences of teething and weaning.

8.2 Auditory and visual experience

Auditory experience has traditionally received less scientific attention than the other experiences enumerated above, perhaps because it has been considered mainly in relation to the development of language and therefore placed at a later developmental stage than that of the first mental levels. On the other hand, recent research in fetal echography has revealed the tremendous importance of this type of experience precisely for the construction of the earliest mental structures (Negri 1993; Imbasciati 1997; Manfredi and Imbasciati 1997; Manfredi, Imbasciati and Ghilardi 1997). Our attempted description of the protomental origins of the mind must therefore begin with a consideration of fetal auditory experience, as the foundation of the first operational levels of the mind system, which therefore determine all its subsequent operational levels.

3 That contribution describes, in particular, symbolization of motor origin, and more specifically ambulatory symbolization; urethral symbolization, which is relevant to psychosexual differentiation (see also Imbasciati 1991); and anal and genital symbolization, all of which are considered from a standpoint completely different from that of traditional psychoanalysis.

As already indicated (see Sections 1.6 and 3.6), these notions could profoundly alter our entire description. However, our present knowledge of fetal psychic life, whether derived from experimental studies or indeed from psychoanalytic mother–fetus observation, is still insufficient for any attempt to reconstruct the first operations of the fetal mind system at information-processing level, let alone in terms of inferences about the nascent subjectivity of the mind (experiences or proto-thoughts).

On the other hand, an approach to protomental life that is, in our view, still valid, and must therefore be complemented by further research, takes as its starting point the neonate, especially if the neo-Kleinian psychoanalytic tradition is integrated with a model of mind – which is psychoanalytic but different from the traditional model – consistent with the present state of the neurosciences (see Chapter 10). It is therefore appropriate for our 'theory of the protomental' to be based, for the time being at least, on the more secure foundation of our knowledge of post-natal life. Our discussion, then, is based on the post-natal period and the vicissitudes of feeding, and invokes the 'breast' and the consequent internal objects. The possibility that a more exhaustive future study of the construction of the fetal mind might invalidate certain aspects of the description given here cannot be ruled out – in particular, the emphasis placed on the structuring value of negative experiences (pain, or the bad object), given that the fetal capacity for protoperceptual recognition suggests, at least logically, that mental engrams stemming from 'good' entities may be more intense and hence of earlier origin.

Finally, though, the value of a theory lies not in a supposed display of facts, but in its operational utility for the scientist – or, in our case, for the mental-health clinician working with mothers and children – in facilitating understanding of those facts and the connections between them.

For this reason, no attempt will be made here to describe in detail how auditory experience, initially in the fetus and later in the neonate, contributes to the construction of successive symbolizations – engrams – organized on the basis of sets of afferences and traces of data from the sense of hearing, and how the use of such experiences determines the structuring of operational levels. We shall merely point out that the preverbal experience of sound comes to form part of a perceptual world that is extremely important for the infant's recognition of reality, and note that preverbal auditory symbolopoiesis has tended to be neglected in studies of the structuring of individual minds.[4]

The attention of clinicians is particularly drawn to the importance of the

4 This concerns not only exploration of the roots of musicality, but also the foundations of non-verbal auditory communication, which is so important for the deciphering of interpersonal verbal communication – in particular, in clinical psychoanalysis.

modulation of the mother's voice in calming an agitated newborn baby: this shows that sound engrams, for a neonate, are tantamount to recognition of the 'good', so that these, when introduced into the gestalt of sets of afferences read on the basis of engrams indicating that they are 'bad' (which is why the baby is agitated), can reverse their sign, transforming the entire complex into a good object. The baby calms down if the mother is able to modulate her voice appropriately; having perhaps initially rejected the nipple, he then becomes capable of recognizing it again. Furthermore, clinical observation shows that the 'good' valuation of auditory sets as described above is most effective when they are accompanied by appropriate tactile and vestibular messages. This can in turn be traced back to fetal experience, in which a combination of auditory, tactile and vestibular afferences has presumably given rise to gestalts that the fetal mind has registered as 'good'.

Visual experience can be described more fully, given that an older scientific tradition exists. The fact that the retinal receptors convey impulses to the brain from the very first days after birth is psychologically irrelevant – indeed, in the language of adult experience, the baby does not actually 'see' anything, because the retinal input cannot yet be 'read'.[5] In other words, the light-based stimuli from external objects activate the retinal receptors and are translated into afferences which reach the brain but are not recognized as having a meaning referable to some external object that has given rise to them; the external object is therefore not 'recognized', even if an accurate optical image of it is formed on the retina. This is an image only in the abstract sense of the physics of lenses; it has no psychic significance, because mental units – fully constituted visual engrams – capable of recognizing, or 'reading', it do not yet exist. Perception – in particular, visual perception – presupposes mental operations, or primitive learning processes (Fantz 1965; Bruner 1968; Vurpillot 1972; Hutt and Hutt 1973; Bower 1974). However, certain phenomena suggest the opposite interpretation, seemingly furnishing evidence of innate perceptual processes determined solely by biological structure. There are particular configurations of stimuli which, in both animals and man, immediately assume the form of perceptual configurations with a well-defined, stable structure distinguishable from others – namely, the so-called elementary gestalts, or configurations, often comprising simple geometrical shapes, which are immediately construed in the act of perception as sets. These have been studied specifically by the gestalt psychology school (see Koffka 1935; Metzger 1954; Kanizsa 1975). The elementary shapes configured as such include, for example, the circle: perception immediately

5 The initial neonatal visual experience may be postulated as somehow analogous to the Ganzfeld phenomenon (the uniform field) as studied by Metzger (1954). This is consistent with the observation of fetal sensitivity to sudden major variations in light intensity.

structures as circles all configurations of stimuli that can be so organized (Arnheim 1970).

According to this view, then, the meaning of these fundamentally elementary 'objects' does not stem from mental operations, but is based on predetermined species-specific cerebral circuits. These objects are considered to be endowed with meaning – however elementary – independently of any learning process.

Our present ignorance of intrauterine learning casts doubt on this conclusion. The phenomena of imprinting, first observed by Lorenz (1965), are well known. Some animals exhibit a clear predisposition to respond to certain figure-related stimuli; however, this response is not a perception, but only a predisposition to a particular kind of learning – namely, imprinting, which depends on the interaction of the animal's mind with the particular external object of which it has had experience (Hinde 1974). Thus, a duckling will follow its mother if it has seen a duck in the period in which it is susceptible to imprinting, but will follow the researcher as if he were its mother if it saw him moving first. Imprinting, which at first seemed an innate response, in fact entails underlying learning processes. Similar arguments can be applied to other apparently innate perceptual responses: such phenomena might be not innate, but simply 'early', and their learning period might have escaped observation. Again, the reference to fetal experience does not necessarily mean that this experience must have concerned the specific sense upon which the response believed to be innate is imprinted after birth.[6]

As to the transformation of afferences into perception, our discussion of the feeding experience distinguished between memory traces in the form of hallucinatory memories and others recognized as belonging to something outside the self (perceptual memory), which can therefore be used as a criterion for the recognition of an external object. This process is not automatic, but depends on the existence of an optimal time lag between the onset of the need and the presentation of the satisfying object (see Section 5.2). If this hypothesis is applied to the experience of visual perception, it is always possible that a substantial interval might elapse between reception of the afference and satisfaction of the need. The sequence considered so far disregarded visual perception and took account only of hunger, the mental processes of exorcism and hallucination, the bottle in the mouth and the final state of well-being. However, the baby may see the bottle lying on the table and still have to wait for a while before feeding, so that the visual sensory stimulus does not immediately and automatically precede satisfaction. The result is a considerable variation in the way this sensory stimulus may become organized in a psychic experience. Whereas the tactile-gustatory

6 Cf. the concept of transmodal perception and learning (Stern 1985).

(oral) afference is readily organized into something meaningful owing to its immediate proximity to satisfaction, an afference more distant in time from the moment when the state of suffering is relieved is likely to be more difficult to transform into a perceptual experience. Hence the retinal input from an object may well assume mental significance only to the extent that the object comes to be related to an experience of need satisfaction, and probably to another internal 'object' already formed in connection with oral experiences.

If the particularity of the visual experience is considered to lie in the integration of the time lapse between afference and satisfaction, this has two implications. The first concerns the greater difficulty of organizing the visual experience into meaningful data, and the second the importance of the sense of sight in the formation of bad objects. A visual image may be said to lend itself more readily than tactile-gustatory images to the causation of persecutory experiences because it may constitute an object that foreshadows satisfaction but does not necessarily keep its promise. The act of looking, for a baby, may involve the experience of something like rage, greed, the wish to take possession of the object, and hence persecution and destruction. Hence, the visual object may be said to stimulate and encourage the formation of the wish, and as such to serve as a useful experience; however, while it seems to hold out the 'promise' of the omnipotence of the wish extending beyond bodily proximity – the eye, being a telereceptor, furnishes knowledge at a distance – it denies immediate satisfaction. The visual object in effect promises, offers itself and at the same time denies itself, in ways that can certainly be metabolized on a persecutory level. The visual object is like a breast that is proffered only to tantalize the baby, and is then immediately withdrawn.

The act of looking – the look – may then be experienced as the vehicle, in fantasy, of the attack on the breast-as-visual-object; its persecutory elaboration is as an external object that can 'look' destructively at the baby. This is in fact the content of many hallucinations of adult psychotics: objects or persons look at the subject with bad intentions and transmit malign influences. The frequent, generally terrifying, visual hallucinations occurring in almost all delusional syndromes can also be interpreted along these lines. Moreover, envy itself has been anthropomorphically portrayed as a person with an evil eye, as in the expressions 'the envious eye' or the 'green eyes of envy'. Angry greed, too, is symbolized in the visual experience: the phrase 'eating up or devouring with the eyes' indicates the experience of vision inspired by an angry, unsatisfied wish.

8.3 The experience of prehension

From about the fourth month of life, following the incipient experience of vision, the organization of the experience of prehension commences. Although the baby was already able to extend its hand and grasp an object, that grasping, which is present even before birth, is purely a reflex action: the baby stretches out his hand and grasps not because he wants to grasp, but because an object has been placed in contact with and stimulated the dermis of the palm of the hand. 'Prehension' proper, by contrast, exists only when the baby grasps – with an initial 'intentionality' – an object he has seen or otherwise perceived. The experience of prehension may therefore be said to become organized when the baby has already acquired the capacity to perceive external objects and is thus in some way capable of 'wishing for' and seeking them. Prehension can in general be regarded as subsequent to and closely bound up with visual experience.

The experience of prehension involves two orders of neurological data that must be organized and transformed into a 'mental' set – namely, proprioceptive afferences on the one hand and motor efferences[7] on the other hand. The former, stemming from the muscular and osteotendinous receptors of the hand, must be processed in order both to be transformed into perception of the external objects that are grasped and to be continuously integrated with the motor efferences that regulate the magnitude and quality of the movements entailed in the prehension of those objects. These efferences thus join up with the afferences to yield a 'knowledge', however rudimentary, of the object taken into the child's hand: the motor program produced in each individual instance is itself constitutive of the perception of the object.

The tactile afferences that always accompany the proprioceptive ones, as well as visual afferences – because the object the baby wishes to grasp has in most cases been 'seen' first – are integrated with the two classes of neurological data. This need for integration presupposes a substantially consolidated capacity of the mind to recognize an object external to the self, and renders hallucination of the object difficult. Whereas an object can be hallucinated on the basis of its visual memory traces or, less easily, of its tactile memory traces, it is much harder to use a proprioceptive trace, because, for one and the same object, this is continuously variable and continuously integrated with a motor program – also contained in a trace – which likewise varies continuously. Hence, the experience of prehension is by its nature oriented toward the recognition of a reality external to the self. Again, this recognition is hugely reinforced by the fact that prehension is an operation that has an actual 'effect' on external reality, serving as it does to reach – indeed, to take – the object that has been recognized. By means of prehension, the baby for the first time succeeds in acting directly on the outside world in order to obtain satisfaction.

However, this capacity for action on and modification of the external world also has a negative aspect: because this experience is more efficacious than its predecessors, it may lead the baby to fantasize an absolute possibility of

7 The afference is the centripetal input in the neural system and the efference its output.

reaching the outside world. In other words, he may come to believe that external reality, even if it is not himself, is at any rate his, so that he is the master of everything and capable of everything. Hence the experience of prehension – precisely because it is effective not only in recognizing of but also in acting on reality – on the one hand stimulates the recognition of external reality, thus placing a limit on the self, but may, on the other hand, make for an experience of omnipotence in relation to the same external reality, except in so far as this is denied by the impossibility of taking the object into the baby's self or of keeping it intact – in other words, by the experience of 'weight' and, in particular, of 'breaking'.

This last experience is particularly important in the organization of the first mental structures; its commencement coincides with the onset of the prehension of objects. This is the stage at which a child tries to grasp everything, and may break things – many things. An object broken is one that disappears as such, thus providing him with a first representation in reality of the effects of aggression, or rather of inappropriate operations, which can be regarded as homologous with destructive fantasies in the mind. The baby had hitherto produced the manifestations described earlier in this book as precursors of rage and envy with confused fantasies constructed on the basis of oral experiences – emptying out, drying up, incorporating, swallowing, hurting inside (the belly) and, in particular, disappearing – or of visual experiences (disappearing, possessing, and also penetrating or devouring); now, however, he has a new and much more effective representation of his attacks – namely, breaking and smashing to pieces. A fantasy of this kind enters the paranoid circuit and therefore symbolizes the possible destructive attack much more tellingly than its predecessors. In the infant's mind, breaking thus comes to embody the prototype of all the effects of his destructive operations and proto-feelings of rage. In so far as the experience of prehension, and hence also of breaking, is construed as a signal of reality and of external objects present in it, which are desirable (B_{e_g}), the representation of the broken object will be experienced as a signal of guilt, which is thus indicative of a bad self, and it will be less easy for it to be metabolized by paranoid-schizoid evacuation. The experience of breaking may therefore facilitate the acquisition of a sense of guilt, felt to belong to the subject himself, and hence a shift toward a more advanced position (the depressive position – see below).

On the other hand, inappropriate prehension, which is precisely what causes objects to be broken most easily, comes to be intimately connected with the destruction of objects; in other words, not only may inappropriate thought operations purely internal to the mind be laden with destructive fantasies (their construction and function being precursors of these fantasies), but also, for the first time, inappropriate operations on the level of action directed toward the outside world may assume the significance of destruction. 'Guilt' then proves to be the same thing as an 'incorrect (or insufficient) action'. Just as the destructive operations directed toward the nascent mental structures involved in the paranoid-schizoid circuits were mental operations intended to avoid the 'worst', or at least what was feared to be the 'worst', as well as unpleasure, but were inappropriate, insufficient or incorrect in terms of affording real satisfaction, now, and to an even greater extent, inadequate operations of prehension of objects become emblematic of their destruction – that is, of the impossibility of obtaining satisfaction with these objects. Hence, the 'incorrectness' identifiable at logical level is equivalent to the 'guilt' identifiable at that of experience. Again, alongside experiences in which the baby breaks something, there are others in which he does not succeed in grasping the object or bringing it close to him, or, if it is an instrument, in using it appropriately, either because it is fixed, too heavy or impossible to take in the hands (as with a liquid), or because the movement is clumsy and insufficient; in all these cases, the baby experiences chagrin and rage, which fuel destructive fantasies toward the object. Infants are often seen to become enraged when they fail to grasp an object or, having grasped it, are unable to use it as they see adults doing; the most common manifestation of this rage is to throw the object away after a few attempts to use it. Hence, the destructive fantasies come, in this way, too, to be linked with, and signified by, motor action; the broken object, as well as the object that has been thrown away and has therefore 'disappeared', is then necessarily seen as the fulfilment of these fantasies, following a motor action that gives rise to guilt – that is, an inappropriate motor action – which is indeed what the child's clumsy prehension in breaking objects amounts to.

An interesting digression would be possible at this point on the child-raising issue of whether an infant should be given all the objects he apparently wishes to grasp, and whether to encourage or try to prevent experiences of breaking he may have with fragile objects. I have discussed these matters in greater detail elsewhere (Imbasciati 1983a), but it is worth pointing out here that the unpleasurable experience of a prehension that fails to grasp the object and/or breaks it may reinforce the child's recognition of the laws of reality, by gradually bringing him to understand that the act of prehension, in order to have appropriate effects on reality and on his own satisfaction, must respect the limits of that reality. After all, if a fragile object is grasped too tightly, it can no longer be possessed because it breaks, or else nothing can be obtained from it because it has not been handled properly. Recognition of the object's fragility – or of its conformation and instrumental possibilities – as a fact of reality means that the infant has worked through his rage at having failed to be totally effective in acting on reality and at having to respect its laws – or, in other words, acknowledging dependence on the external world and on the nature of external objects, with all their absolute and compelling demands. From this point of view, it may be useful to allow the infant to have the experience of breaking. On the other hand, this experience, if excessive, might plunge the infant into despair and confront him with a sense of impotence with regard to his own aggression. These emotional constants, which characterize any over-permissive upbringing, may hinder the infant's ability to distinguish his inadequacy and errors from inappropriate actions to which more guilt attaches – that is, those directly linked to his aggression. That is a distinction he will make later, on the basis of his 'hope' of becoming capable in spite of his present incapability.

Hence, one and the same experience – that of breaking – may be used both negatively and positively by the child's mind; the outcome will depend on how he deals with his 'guilt'. This will be discussed in the following chapters. Possible child-raising issues cannot therefore be considered in isolation from the relational context in which the mind processes a given experience.

8.4 Teething and weaning

Psychic development takes place in two directions: the progressive structuring of functions and the acquisition of a growing number of engrams which these permit. The functions are structured on the basis of the first operations the mind is able to perform. These have been described in terms of their greater or lesser appropriateness ('correct/incorrect operations') for seeking actual satisfaction in external reality and in the recognition of that reality, as well as in internal reality – that is, in the development of the mind itself toward more effective operations. The first mental operations will be the starting point for the successive development of all the various 'functions' that can be identified in the adult mind – both those devoted to knowledge of the external world and those whereby the mind preserves its own self-consistency, or self-knowledge, however limited and variable this may be in individual consciousness. On the other hand, mental operations are possible only if the mind possesses material with which to work – namely, mental contents. We have attempted to describe the formation of protomental contents in terms of primitive aggregations of traces of afferences and of their signifying capacity; we have referred in this connection to mental objects, by which we mean both the first internal objects (which we compared with those identified by Kleinian theory) and all the subsequent objects that gradually come to be manifested in fantasies, and hence in representations that match objects existing in reality. The signifying capacity of mental objects, from the most primitive ones on, results from the quality of operations: for instance, we described how different operations may impart different meanings (external or internal, good or bad, belonging to the self or 'other' than self) to the same sets of afferences. So the entire subsequent divergence of the first internal objects into a wide variety of fantasies and more and more discriminating symbols, up to the stage of the sophistication and realism characteristic of the mental engrams used in adult cognitive processes, depends on the quality of the operations the mind is able to perform. These, however, are in turn influenced by the quality, richness, discriminating capacity and degree of conformity to reality presented by the contents.

In a feedback system of this kind, development results from the use of experience, which, for an infant, centres on his own bodily parts. The innovations that gradually promote the development of his mental structures therefore stem from the possibility of attempting bodily modalities arising out of the use of various, successive parts of his own body. Corporeity, and hence bodily relationships, constitute the foundation of the development of the 'spirit', understood not in terms of a mind–body dichotomy but more generally as the possibility of knowledge (Imbasciati 1978b). The manner in which experiences will be used will of course depend on the structures that have previously formed. The outcome of one and the same experience will therefore differ according to the system of structures by which it is processed. Hence, at every instant of our lives, we are conditioned by our entire previous internal biography. That is why it is so difficult to correct situations in which the whole of an individual's prior experience has structured certain patterns of mental operations, which have become embodied in actual structures. An example is the psychotherapy of severe psychosis or mental retardation (Searles 1965; Zapparoli 1979). In these and other forms, the system of structures established in the first period of personal development – whether we call it the apperceptive mass or the personality – will greatly limit the freedom of the individual concerned (Imbasciati 1978a) and his possibility of change.

The first bodily models presented for cognitive development are oral. In this experience, which we have attempted to describe in logical rather than representational terms, concentrating more on the development of structures (operations) than on symbols (contents), the appearance of the teeth is a fundamental, innovative event. In fact, the relevant 'phase' precedes the actual appearance of the teeth because, a few months before, the baby is already pressing his jaws together and making his first attempts at 'biting'. However, the significance of the experience peaks when biting can be carried out with the teeth. For convenience, we shall describe the importance of this phase with reference to teething, even though an experience of this kind is prior to the eruption of the teeth. Again, the experience of teething, like that of weaning, is relevant to research on the formation of the first mental structures, and both of these areas have been quite thoroughly investigated by psychoanalysis, so that more relevant data and explanatory hypotheses are available.

The experience of teething, being unpleasurable, fundamentally changes the mind's potential for structuring certain operations. The cutting of the first teeth gives rise to a sensation of discomfort in the mouth that leads the baby to press his jaws together and to touch the inflamed area with the tongue, but this affords no relief. A tooth is very sharp; it is a cutting object that hurts and sticks out in the mouth, in the middle of a part of the body that represents the source of operations and is experienced as the centre of the self. Teething is a powerful and telling demonstration – an experience of reality – of the fact that the wish to eliminate unpleasure does not suffice to bring about change.

By the preceding mental operations, the infant will have come to observe that his immediate impulse to eliminate the unpleasure calls for correction. When he was hungry and remained unsatisfied, it was easy for him to anaesthetize his mind and forget the hunger, but, with the eruption of the teeth, he cannot readily fail to register the presence of such a strong – and biologically significant (see Section 4.2) – stimulus and of the consequent massive and intense afferences. The infant may be imagined to experience great irritation and furious rage: it is impossible for his mind to inhibit (autotomize) the transformation of afferences into psychic elements, because the painful afferences cannot be eliminated. The only remaining possibility is to project this evil; the result is the constitution of a persecutory breast with teeth, which not only wants to devour but also bites and tears. The persecutory phantasm is enriched with cruel attributes, and the child will fear the breast and his own desire to bite. This desire is experienced, turned back on the infant himself and heightened, in all the terror of the paranoid-schizoid position: the persecutory breast now assumes concrete form in a range of intense fantasies suitable for 'representing' the unpleasurable experience of teething and, in particular, of states of rage and aggression – that is, the destructiveness which, having undergone paranoid-schizoid metabolization, rebounds in heightened form on the subject himself and attacks his good internal objects – namely, thoughts. Since the bite the child wishes to inflict on his object owing to the stimulus of hunger, of the teeth and of rage transforms that object into a persecutory one, it can readily be imagined how the projective dynamics give rise to experiences of a more immediate destructiveness. The experience of the child's own destructiveness is now based on symbols with the concrete aspect of almost tangible pain; the rage becomes an impulse to tear to pieces to the point of drawing blood, just as the child's gums bleed if he presses them together too hard.

If the baby already has teeth and still sucks at the breast (this is less usual today, owing to early weaning), this proves to be an important experience in which the internal and external worlds are made good: he discovers in the nipple an entity which, if bitten, does not retaliate by inflicting pain, but which must nevertheless be respected – that is, must not be attacked. Nipple-biting occurs well before the appearance of teeth, but has the same effect, albeit less intense. Since the baby does not feel pain when biting the breast, he bites it – but the nipple is withdrawn. The baby then feels frustration and chagrin at the failure of his omnipotence; this gives rise to rage and aggression, so that he may hate the breast and in turn fear a counter-attack from it. However, in so far as the mother is able to accept the projective identifications inherent in the attack, and to convey a good message back to the baby, she is demonstrating to him that the breast does not exact revenge – that is to say, that there is no persecution, but only a real breast, which, however, must be respected. In other words, in this experience of feeding a child who is beginning to bite her, the mother has an opportunity of interrupting the paranoid-schizoid circuits of the infant's mind and of teaching him 'reality'; this reality is a breast that cannot be bitten and must be respected – that is, loved – and recognized as having requirements of its own, if it is to be preserved for the infant's good. In this way, he will learn the rudiments of a relationship with a living object, as the foundation of his future capacity for interpersonal relationships – in other words, the capacity to love. At the same time, he can begin to get to know what is outside himself and to familiarize himself with the characteristics of these objects (or persons), their nature and the (natural) laws to which they are subject. Every breast-feeding mother teaches her child not to bite her in so far as there is an exchange of unequivocal communications, enabling the child to understand that the nipple is not bitten even if he does not feel pain, and that this ban on biting it does not signify revenge. In this context, mother–child communication involves the capacity to accept the paranoid aggression of the child's mind without returning it – that is, without the mother herself responding by a counter-attack directed toward elimination of any pain inflicted on her by the child's action. Only in this way can the mother break into the paranoid-schizoid circuits and 'teach' her child reality – that is, teach him to know and to think. It is thus essential, in the mother–infant communication, for the mother's mind to understand paranoid-schizoid functioning and continuously to 'translate' it for the infant into a different form – the depressive or realistic form, as we shall see later – so that he can 'learn' it.[8] The material gathered by the technique of infant observation (Bick 1964) and discussed by Rosenbluth, Harris, Osborne and O'Shaughnessy (1969) includes a wealth of similar situations.

Biting constitutes the basis of the experience of something bad that can happen outside the infant's self – that is, to someone else – which is 'bad' even if the subject does not directly feel pain. It is therefore an experience that underlies the knowledge of an 'Other' as a person. The infant's emerging interest in ensuring that the other-than-self does not suffer lays the foundations for his seeing this 'Other' as a separate person. In this way, the conditions gradually come into being for a respectful relationship with the Other, which is affectionate in that it involves a sense of unison with the Other, so that the infant experiences bad things that affect the Other as if they affected himself, and is concerned for the Other's well-being. The relevant emerging experience of reality obliges the individual to take an interest in the

8 The dynamics described here presuppose breast-feeding; the situation may well be different in the case of artificial feeding, when the baby is in a different form of bodily contact with the mother and what he has in his mouth does not possess the 'animate' characteristics assumed above. These issues constitute an as yet largely unexplored and complex field of study, which I have discussed in more detail elsewhere (Imbasciati 1983a, 1990).

well-being of the nipple. This of course presupposes an interest in the subject's own well-being ('if I bite it, I shall no longer have it to suck'), but this personal experience causes the infant to realize that the object does not belong to him, that it has its own requirements and that it is in his interests to respect these, because his own well-being depends on that of the nipple; it is this experience that constitutes the basis of a mental model of affection and love. There is thus a transition from self-advantage to coexistence.

This situation too is fraught with potential problems. For instance, if the mother is suffering from rhagades, this will impair and alter her communication with her child.[9] In this case, leaving the nipple in the child's mouth is already a problem for the mother; she therefore cannot allow him to have even the slightest experience of biting, which, if tolerated, would convey a signal teaching him not to bite. Such complications may induce the mother to wean the child just when he is already undergoing the frustrating experience of teething, so that the two experiences are superimposed and the nascent mind is presented with a threefold task, centring on recognition of the Other as a separate person on whom the subject depends, on concern for the Other's well-being and on the subject's own painful and frustrating experience. As Winnicott (1957: 68) writes, 'weaning is not only getting a baby to take other foods, or to use a cup, or to feed actively using the hands. It includes the gradual process of disillusionment, which is part of the parents' task'. The reality that must be accepted by virtue of this disillusionment is primarily interpersonal reality, given that the initial situation – feeding – in which the world has impinged on the child and adapted to his needs now enables him to respond gradually to the demands of his 'environment', with the result of genuine baby–mother collaboration. In this context, weaning is accompanied by a substantial broadening of the child's experience.

Rosenbluth, Harris, Osborne and O'Shaughnessy (1969) point out that weaning, as an organizing or prototypical experience, can be regarded as already occurring in the first months of life, in so far as the maternal breast or the bottle is at the child's disposal at certain times only, whereas he has to do without it for most of the day. In English, the verb 'to wean' originally meant 'to accustom or habituate', and the baby must indeed become accustomed to the loss of feeding contact[10] with the mother between feeds; he will then have

9 It is unclear how far rhagades may be due to causes in the mother only (and her skin metabolism) or to a particular relationship established between her and the child, whereby his sucking behaviour has been modified, thus psychosomatically influencing her skin metabolism.

10 The Italian verb *svezzare* literally means 'to dehabituate', but implies the enforced loss of something pleasurable. The etymological meaning is 'to remove the *vezzo* [habit]'. A *vezzo* signifies something pleasurable, to which the baby has become accustomed and which he finds it hard to give up. Children are often 'naughty' because they want to keep their *vezzi*.

had the experience of doing without maternal contact. Weaning also signifies becoming accustomed to a new kind of feeding – as well as to a different form of contact and hence of interpersonal relating, involving different qualities and feelings which, in the normal course of events, reinforce the nascent mental structures that make up the subject's own self. That is what Winnicott (1957: 66) implies in his paradoxical comment that, in 'weaning, the aim really is to use the baby's developing ability to get rid of things, and to let the loss of the breast be not just simply a chance affair'. As already noted several times, this calls for attunement of the mother with her child, so that she can administer the inevitable frustrations of development in gradual doses, tempering them by her own capacity to understand love and hate, attachment and disappointment, and the baby's rage and sadness. All is set fair for the baby if he is weaned when he has already discovered that, behind the breast or bottle, there exists a mother as a whole person.

Weaning thus performs an important function in the process of getting to know reality and involves mental work on the part of both members of the mother–child couple, bound up with experiences of separation, the sense of loss, and longing. In the words of Harris (1969: Section 24)[11]:

> For the mother who has been breast-feeding her baby, weaning is a change that affects her intimately. Breast-feeding has been an enjoyable experience for her, giving her leisure to satisfy and promote the growth of her baby in such an intimate way and to identify with the baby's satisfaction. Weaning, therefore, is very much a giving-up process, a facing of separation for her too. It involves a little mourning, for her as well as for the baby [. . .].

8.5 From the body to 'fantasy'

We have described how different sensory – or at any rate bodily – experiences are organized in such a way as to give rise to a gradually expanding range of ever more complex engrams. At a certain level in this progression, these engrams correspond to what is described in the Kleinian literature as 'phantasy' or 'fantasy'. This has to do with relations between internal objects, which in effect constitute scenarios that may be variously represented by (in fact unrealistic) images of bodily parts and functions (such as eating, devouring, swallowing or burning) or bodily zones transfigured by projections, and which make up the infant's internal world. For these descriptions, Klein used the spelling 'phantasy' rather than the common form 'fantasy', to indicate that these were not fantasies in the more mature and conscious sense of

11 [Translator's note: The pages of Harris 1969 are not numbered.]

daydreams.[12] Since Klein, the spelling 'fantasy' has predominated in the literature, especially of American origin, the original sense of the internal world with its internal objects being taken for granted. Our theory describes 'proto-' mental events that have not yet attained, but are precursors of, Klein's 'phantasy'. In this book, the continuum extending from the first engrams to the Kleinian internal objects has been repeatedly emphasized. The term 'fantasy' could be used to denote this continuum, bearing in mind the comment in Laplanche and Pontalis (1964) (see footnote 12). However, 'fantasy' can be understood only in the framework of a Klein-derived conceptualization (Isaacs 1948; Segal 1964a). For this reason, to take account of the continuity between the most primitive engrams and the Kleinian internal objects that succeed them, the word '*fantasma*' has sometimes been used in the Italian version of this book.[13] We have likewise stressed that the concept of the internal object can also be understood in a wider sense, as proposed in Chapters 3 and 4 of this book. The term as employed here thus covers a wide range of mental events and configurations. Similarly, 'thought' is used here for the entire process extending from the most elementary protomental operations to the complex, logical-rational constructions that characterize the various aspects of thought as more strictly defined in philosophical language and common usage, even embracing the collective elements called 'culture'. As we have attempted to show above, the formation of fantasies is made possible by the child's progressive use of his sensory apparatuses,[14] centring on the oral zone. The transformation of this zone with teething changes the mental representations and links them in a more varied and complex manner. This leads to a larger quantity of symbols and the possibility of using them

12 [Translator's note: The spelling 'fantasy' is used almost universally in the USA, while both spellings are current in the UK. 'Fantasy' is used throughout this translation except where 'phantasy' occurs in quotations. 'Phantasy' is also the form favoured in the *Standard Edition* of Freud. Laplanche and Pontalis (1967: 318) write: 'In her article of 1948, "The Nature and Function of Phantasy", Susan Isaacs proposes that the two alternative spellings *fantasy* and *phantasy* should be used to denote "conscious daydreams, fictions and so on" and "the primary content of unconscious mental processes" respectively. Isaacs feels that such an innovation in psycho-analytic terminology would be consistent with Freud's thought. In our view, however, the suggested distinction does not do justice to the complexity of Freud's views. In any case it would lead to problems of translation: if, for every occurrence of "*Phantasie*" in Freud's writings, a choice had to be made between "*ph*antasy" and "*f*antasy", the door would be open to the most arbitrary of interpretations.']

13 [Translator's note: The Italian word is retained here because the author is using it in a sense different from 'fantasy' (or 'phantasy'). It has sometimes been necessary to render it by 'phantasm' (in the sense of a ghost).]

14 Mastery of the organs results from the mental organization of prior experiences with them. Biological maturation is a necessary but not a sufficient condition for this; there can be no mastery without prior experience and the capacity to use it, as are obtainable with the organs in a mental organization.

with greater precision for knowing reality – not only external reality, but also internal reality in its destructive aspects (sadism). This applies in particular to the possibility of metabolizing counter-persecutory aggression, as well as to the moulding of that aggression with respect to the reality that is most likely, for the relevant sense, to trigger aggression and persecution.

As stated in Section 8.2, the experiences of vision and prehension interact with the fundamental oral type of fantasies and, whereas the former tend to channel the symbolization process toward the discrimination of external reality, they are nevertheless modulated in accordance with the levels of the paranoid-schizoid position. The visual external object, to the extent that it begins to be perceived, constitutes a reality test because it stimulates the wish independently of need, while at the same time encouraging the desire to seize everything that is seen or to destroy it if unattainable. These two aspects are reflected in common expressions such as 'envious eyes' or 'gobbling up with the eyes', which signify that the visual organ lends itself to symbolizing the greedy wish to incorporate objects and the envious wish to attack them. By contrast, ear symbolism, for example, while also involving incorporation, lacks the stamp of aggression, as in the phrase 'to drink in someone's words'. Similar considerations apply to experiences of prehension:[15] these lead to the recognition of reality because they are the first operations (subsequently to be integrated with locomotor schemata) suitable for reaching the object; however, they also make for an experience of omnipotence, since they induce the infant to believe that, if things can be grasped, they are a kind of appendage to himself and therefore always at his disposal.

Along the same lines as the above description of the mental organization stemming from auditory, visual, prehensorial and dental-masticatory experience, a similar account could be given of experiences of respiration, olfaction and taste, vestibular experiences and, in particular, the important sphere of motor experience – especially locomotion. An analogous redescription of urethral and anal experience would also be possible, revealing them in a different light from that of the psychoanalytic tradition, within the logical-cognitive model of the various levels of protomental operations gradually constructed through the exercise of these bodily apparatuses (Imbasciati 1983b). In the same way, the construction of the sexual dimension and of genitality could be reformulated in terms differing from those of traditional psychoanalysis. Each of these sectors of experience revolves around a particular bodily zone or function; others, in addition to those enumerated or described elsewhere (Imbasciati 1983a), could also be investigated. For our present purposes, each may be seen as giving rise to a specific kind of

15 Analogous verbal forms (e.g. com*prehension*, or *grasping* a concept) are remote echoes of the links between prehension and the recognition of reality.

symbolization, superimposed on others, which are determined by their predecessors and in turn determine their successors, and which, in particular, interweave with all the others to form a progressive network of ever more complex symbolization processes that will allow the development of the adult mental capacities, the emergence of consciousness, and the mature individual's capacity for abstraction. The route is a laborious one: a 'symbolic series' interwoven with the others in 'chains of symbolization' (ibid.) so as to form the entire symbolopoietic development may be deemed to emerge from each type of bodily experience. The general pattern is illustrated in Chapters 9 and 10 (see Section 9.6 and Figure 10.1).

Let us now continue our comparison of the classical psychoanalytic description with our explanatory theory of the construction of cognitive processes. The symbolization processes may be said to lie on a continuum extending from the 'internal objects' of psychoanalysis to the concepts described by psychology as representations of external objects or, from a different point of view, categories used for perceiving and judging events (i.e. for knowing) – that is, cognitive 'capacities'.

In traditional psychology, the term 'internal object', while little used, can denote our mental image of an object that exists in reality. This implies that the image and the object coincide, on the consciousness-based assumption that the image conforms to the shape of the external object. In the psychoanalytic view, however, an adult's images of objects existing in the external world result from a process of linking and imaginative creation whose origins lie in the individual's first mental experiences. These persist in the structure of the adult mind, forming different levels that regulate the unconscious processes of thought:[16] the internal objects in the specific psychoanalytic sense have been described in terms of these levels. In our description, the term 'internal object' is used in a broad sense, embracing, first, a mental entity more primitive than the internal objects of psychoanalysis; second, the psychoanalytic internal objects; third, the way in which both types of object provide the neonate with his own particular vision of the world – that is, a protorepresentation – and, last, the manner in which all the previous entities gradually give rise to increasingly mature engrams (with their associated increasingly mature operational levels), culminating in reality-matching representations proper.

An outline can now be given of the progressively more complex internal configurations whereby the most primitive symbolizations lead on to more

16 The intensity of unconscious thought processes is emphasized throughout the work of Bion (1962, 1963, 1965, 1967). Bion may also be credited with the origination of a model in which the functioning of thought is conceived as the simultaneous activation of all the levels that progressively begin to function in an individual's developmental history and are progressively interlinked. We normally observe only the final, surface effect of this simultaneous activation or reactivation – namely, conscious thought.

developed ones – for example, to representations capable of genuinely 'representing' reality in such a way that the individual can function effectively within it.

These 'representations', in adults, match reality, or rather, they match reality 'sufficiently' accurately. In fact, there is never a simple, faithful reproduction of reality: the discrepancy between unconscious vicissitudes and reality-oriented conscious thought revealed by psychoanalytic research may be likened to that other discrepancy, emphasized by projective psychology, between perception and 'objectivity' (cf. apperceptive distortion – Imbasciati 1967). This disposes once and for all of the conception of early psychology with its implicit assumption of the constancy hypothesis (see Section 2.3). The concept of a 'sufficiently accurate match' applies not only to the capacity for representation but also to all the other capacities gradually acquired by the mind, and constitutes a good description of the continuum extending from the first processes of symbolization, which are so far removed from reality that they seem to be exclusively endogenous, to their later counterparts, which are increasingly suitable for knowledge.

The internal object is described by psychoanalysis as an affective object. This means that psychoanalysts use the term 'cognitive' in the more traditional, restricted sense of logical-rational processes. However, this use is reductive, for the affects too are basic cognitive schemata (cf. Plutchik 1980); furthermore, Money-Kyrle (1968) long ago recognized the cognitive nature of the internal objects as equivalent to the child's capacity for representation. Since the so-called affective processes therefore serve the purpose of cognition, they, too, must be ascribed to the cognitive sphere, which in turn cannot be confined to the logical-rational processes, let alone the conscious ones. In other words, the affect–cognition dichotomy is spurious and reflects the adult mind's difficulty in understanding protomental processes (Imbasciati 1991). The primitive internal object is so difficult to describe because, being preverbal, it is unverbalizable. It depends on archaic experiences organized by laws different from those which, by adaptive logic, regulate the representational adjustment of the adult mind so that it can interact with reality. For example, in a child, the contents of the mind and the associated mental functions are inseparable from each other (see Section 7.4), whereas adults, on the whole, when reflecting on what they have thought, are able to distinguish between the product of their thought and their attitude toward this object of thought. A child, however, lacks this capacity.[17] Until the internal objects come to be organized in engrams corresponding to reality, the modalities of thought are

17 Likewise, there are some adult subjects who can distinguish the existence of what they perceive from the fact that they are perceiving it in that way while being aware that it could be perceived differently (cf. the discussion of alexithymia in Imbasciati 2004b).

governed by a particular logic based on primitive mental operations. For this reason, any attempt to describe such objects in adult language necessarily calls for circumlocutions and figurative concepts whereby such confused and archaic entities can be linked together and described more appropriately.

In the construction of symbolization processes, sensory experiences are organized by the successive operational levels built up in the mind in such a way that the primal internal objects are linked by subsequent processes of transformation and become other objects, which are likewise internal but more differentiated, with the eventual result of images and thoughts that match the characteristics of the external world sufficiently accurately. In other words, there is a continuum in the transformation of utterly primitive proto-objects into internal objects corresponding to those described by psychoanalysis and to still other objects that correspond to what used to be called 'representations' – however, all of these are representational in nature. For this reason, the term 'internal object' is used here in a broad sense.

The process of symbolization is described in the psychoanalytic tradition[18] – in particular, by the Kleinian school (Klein 1930; Segal 1957, 1978) – in terms of feelings (hate, love, guilt, envy, greed or anxiety) and bodily images (the breast, mouth, teeth, anus, faeces, urine, etc.), as well as by means of accepted psychoanalytic concepts denoting processes (defence, impulse, drive or reparation). In our description, which is, in our view, more detailed, these notions are referred to the progressive interlinking of traces of afferences and the progressive recording of functional traces. The function of the process of reparation, as already noted and as will be emphasized again below, is described here as the restructuring of bad-object engrams contained in the persecutory-paranoid-schizoid circuits so as to form good-object engrams that can be retained and not autotomized, thus constituting symbols effective for the purpose of knowledge. It is therefore a matter of repairing symbols that would otherwise serve neither for the knowledge of reality nor for the

18 With regard to the psychoanalytic definition of a symbol, Money-Kyrle (1961: 70f.) draws attention to the difficulty of clarifying the relationship between the psychoanalytic and ordinary senses: 'As distinct from those objects, later to become symbols in the analytic sense, which are for a period equated with what will later become their objects, a symbol in the ordinary sense, for example an idea or a word, is quite distinct from, and never mistaken for, the object. Perhaps what is confusing about the psycho-analytic symbol is that it has characteristics of both and represents an intermediate stage between the two. Intellectually, it is no longer equated with its object but it arouses emotions which may be quite as, if not more, intense. [. . .] Perhaps therefore we should recognise yet another transitional stage in which "symbols" arousing persecutory feelings are still equated with their objects, whereas those arousing depressive feelings have begun to symbolise them in the analytic sense. If so, true symbols both in the analytic and the ordinary sense arise when they begin to *stand for* something at first thought of as missing and irreparably lost, and later as only temporarily absent. They may do this either unconsciously or consciously.'

construction of further symbolizations. The 'goodness' of repaired symbols, which the Kleinian theory describes in terms of love, is reconceptualized by our theory in operational terms, in which symbols are good if they serve the purpose of development – that is, of relating effectively to the outside world – so that needs can be satisfied. From this point of view, 'love' is a later entity, connected with the transformation of needs into wishes – and with a wish linked to the possibility of obtaining something by means of an effective mind. This, then, is a love inherent in mental development itself, and in knowledge. The human mind thus contains in its specific development a need, and later a wish, for development itself. The necessity of continuous development is inherent in the dialectic of the symbolization process in the human species. Hence, this process proves to be not a mere extension of the mind for biological survival (Imbasciati 1979c), but also a continuous development directed toward progressive 'mental survival'.

This chapter is intended as an introduction to our discussion of how the later operations of the 'depressive situation' make use of much richer and more complex mental products than their predecessors – namely, the internal products described in the Kleinian psychoanalytic tradition as fantasies. For convenience, our examples will continue to use the terminology and symbols of oral experience – such as the 'breast' – but it should be borne in mind that the 'good-breast internal object' of the depressive position is a much more sophisticated mental product than the internal object of the previous operational levels, which are made up of very different sets of traces of afferences and other engrams.

Chapter 9

The depressive position and learning to know

9.1 Antecedents of the 'sense of guilt'

As stated at the end of Chapter 6, the mind system can acquire the capacity to perform operations whereby it can distinguish representations or protorepresentations of external objects (which thus begin to be *accepted* into it by more specifically *perceptual* processes) from processes (feelings or protothoughts) of internal origin that relate to 'those' objects. In other words, the external objects are distinguished from the subject's own internal processes (thoughts) so that the latter are experienced as constituting an initial self, which is no longer confused with the rest of the world. As a result, the subject's own impulses of hate, rage, envy and aggression – that is, the 'bad' – are also felt to belong to the subject himself. So the presence of a bad self inside the subject must be tolerated. This incipient bad self, which becomes increasingly ramified through the use of fantasying based on functional models derived from bodily experience (see Chapter 8), assumes the form of the possible experience of attacking 'the good objects', chief among which are the subject's own nascent mental capacities.[1] This is the threshold of what used to be called the sense of guilt – or rather, of internal events describable as *antecedents of an experience of guilt* (cf. Section 7.4). It is the threshold of Klein's 'depressive position', which she saw as the source of the possibility of good object relations (the capacity for love) and, on the basis of these, of good psychic development. In a further theoretical development of these ideas, the Bion school regards this position as the locus of the origin of 'thought', considered as a positive and decisive shift away from the evacuative psychotic processes (Ps $\rightarrow$ D) of the paranoid-schizoid situation. Let us now attempt to describe the depressive position in terms of our model – that is, by identifying the 'cognitive operations' inherent in it.

1 In information-processing terms, a signal arises indicating the possibility that autotomic processes ('negative thought') might erase the adaptive mental processes ('positive thought') – that is, the mind's nascent operational capacities.

Freud originally described the (unconscious) sense of guilt as a transgression of the introjected parental rules, bound up with the origins of the superego. This initial formulation was espoused more widely with the spread of psychoanalytic ideas in the fields of sociology and education. As a result, theories (some of them 'scientific', although outside the specific discipline of psychoanalysis) were propounded for decades on the repressive effects of an excessively severe upbringing, which was held to be responsible for – precisely – guilt feelings rooted in the deep structure of the personality. Moreover, the sense of guilt was considered to be a potential source of pathology or, at any rate, an obstacle to good psychic development. Typical examples of this position can be found in Alexander (1956) and Marcuse (1955, 1960).

Although Freud at first attributed the sense of guilt, often associated with sexual psychopathology, to a repressive upbringing and the introjection of over-severe parental rules, in his later writings (1923b, e, 1924d, 1925j, 1933a [1932]) he found that the sense of guilt and the superego can be traced back to the child's fear of losing his parents' love, and that this, rather than the imposition of an authoritarian upbringing, causes him to introject the parental rules. In some late passages, Freud explained that this 'fear' has to do not with circumstances in which the child was actually deprived, or at risk of being deprived, of parental love, but with something internal connected with a more general fear of losing love. Following up Freud's observation that the severity of the superego does not always reflect that of the treatment meted out to a child, Grinberg (1976: 51) notes that a 'child brought up indulgently may also develop a strict moral conscience; excessively kind and indulgent parents may provoke in the child the formation of too strict a superego'. Freud's formulation opened the way to subsequent Kleinian developments, in which the term 'sense of guilt' referred to a much more primitive process than the more advanced ones it had previously denoted. These developments in fact constitute a different conceptualization of a different process, or rather different processes and phases in the course of mental development, to which the same term – guilt, or the sense of guilt – is applied.

The Kleinians took the psychoanalytic study of the sense of guilt back to more remote origins than those considered by Freud: his intuition that the sense of guilt is connected with loss of love is explained by the Kleinian school in terms of the stages of life and mental processes when the conception of love and its possible loss first arises in the mind – that is, in the vicissitudes and dynamics of the first objects. Hence, the sense of guilt originates, logically enough, in the protomental processes. What is feared is not a realistic loss of love, as Freud seems to have believed, but the loss of something internal that for the child represents love, out of which his entire capacity for love arose; it is therefore a loss of primary internal objects. What is feared in situations of guilt is the loss of something good that has been built up inside – of some good internal object felt to belong, as an incipient self, to the subject. The first internal objects (as broadly defined) which the proto-mind is afraid of losing are actually its own nascent structures, as described in our theory. The sense of guilt thus stems from a proto-experience of the possible occurrence within the mind of a catastrophe, a destruction of mental processes, or rather a destruction of the structures (engrams, internal objects and fantasies) that constitute its primary aspect – so to speak, the raw material, the precondition and promise of all growth.

In the Kleinian formulation, the sense of guilt is related to the child's first internal objects and aggressive fantasies, which, as it were, constitute dramatic presentations of mental states of rage, hate, violence and envy toward the first object of his experience – the breast – and its subsequent extensions in the person of the mother. The attack on the objects is held to turn them into objects ruined by envy and aggression, so that nuclei of damaged objects, which the infant would like not to have ruined, form in his mind. Klein (in my view, seeing the situation in adult terms) saw this state as a feeling and was thereby led to describe it as a sense of guilt. These damaged objects in the child's mind take the form of a warning of the possibility of aggression against its own love objects – that is, as a sense of guilt (Fornari 1976). However, their function is not only of warning but also of aggression, in so far as the attacked object is easily transformed into a persecutory object: in the dynamic of the sense of guilt, there also develops the dimension of accusation and oppression, which are not readily tolerated and are therefore projected, giving rise to a counter-attack, and hence aggression. The prevalence of persecution in this process – that is, the fact that the sense of guilt is projected – is typical of the paranoid-schizoid situation. But if the sense of guilt remains in the mind as an experience of the subject's own, of having destroyed or being capable of destroying something, it constitutes the characteristic state to be described below – namely, the depressive situation.

In my view, Klein's formulation emphasizes the external object more than its internal counterpart: although she refers to an attack on the good internal objects, she sometimes appears to equate these with the external objects from which they originate, independently of projective processes. The good breast, for her, is internal because a good external breast has been introjected. In other words, she seems to be saying that the attack on the good objects takes place because the internal objects stand for a destruction of the external objects, so that their destruction is feared. Furthermore, her formulation on the level of feelings, within an overall adultomorphic conception of the infant mind, makes it difficult to describe the internal world in representational terms, and hence to formulate the primary sense of guilt more clearly in terms of protomental events. On the basis of Bion's descriptions of the formation of thought (Bion 1962, 1967), the mind can now be said to fear the destruction of its first internal objects independently of their ability to *stand for* concrete external objects, but instead in relation to their internal function. These objects are, after all, the vehicle for *representation*, in whatever form, of the entire external world, and they underlie the capacity to establish a relationship with the world – that is, knowledge. They are the very foundation of the mind. For this reason, the mind system, even if it autotomizes itself at the primitive stage, fears this as the most catastrophically destructive operation possible. In other words, the sense of guilt can be regarded as a signal, which the system must learn to use, of the possibility of self-destruction of the system's own functions.[2] In this primal context, the sense of guilt plays a fundamental part in the protomental cognitive processes.

In describing the construction of mental operational schemata, we have shown how, in the transition from hallucinatory operations directed toward the avoidance of pain to effective operations carried out on reality, the individual experiences modes of mental functioning that lead to a kind of possible destructuring and loss of his own mental organization. The first 'incorrect' operation arising has self-destructive effects, in which a constitutive element of the persecutory aspect of guilt can already be discerned, because the possible recognition of having destroyed 'something inside' can be conceived by the subject only as an emptying out. A hungry baby, in order to deny his sense of hunger, empties himself out internally and in so doing also gets rid of something mental – namely, his mental experience of hunger that was in the process of formation. This merely accentuates his state of need. In favourable circumstances, his mind may begin to operate realistically, recovering the signal of hunger and using it to construct a mental experience that leads him to satisfy it. Alternatively, he may persist in his self-destructive circuit, for instance by destroying the nascent mental experience of the need to suck in the event of hunger; instead of sucking, he may then even refuse the breast. By doing so, however, he splits off from himself his initial experience of hunger, thus emptying himself of something mental. By attempting also to split off from himself this experience of being emptied out and destroyed, he projects it in persecutory fashion on to something external to himself, which becomes the persecutory image of a bad breast that wants to empty him out and eat him up. In this way, the mind feels 'devoured' by the sense of guilt.[3] By splitting off and evacuating from itself its own

2 Many authors associate the sense of guilt with the death 'instinct', of which some regard it as the direct expression. In fact, however, the *signal* of guilt is the most elementary defence against death and self-destruction.

3 At this primitive level, the sense of guilt, or rather guilt-related anxiety, does not yet take the form of an urge to remedy the situation, but appears as something monstrous and terrifying, made up of the following successive equations: (1) badness-hunger = (2) evacuation of badness-hunger = (3) emptying out of incipient mental content = (4) being devoured = (5) being devoured by having destroyed the mind through emptying it out = (6) the devoured incipient internal breast constituted by the nascent mental experience + projection = (7) the persecutory devouring bad breast + reintrojection = (8) the intolerable mental experience of extreme anxiety, which thus devours itself, attempting to destroy itself again by further paranoid-schizoid mechanisms. This schematic outline, which is useful as a brief description

experience of emptying itself out or somehow destroying itself, the mind empties out and destroys itself still further.

The particular painfulness of the mature sense of guilt can thus be explained by these origins: as a signal of destructiveness, the sense of guilt may take the form of a tormenting and persecutory complex of uncertainties and contradictions, the matrix of an accusation of destruction that compels the mind to destroy itself even more in order to avoid it (see Section 5.4). This mechanism shows that the experience of guilt has a cognitive, information-processing aspect, signalling as it does that an incorrect, antirealistic operation has been performed in the mind. In cognitive terms, guilt then appears in the form of 'error'. At protomental level, however, the system is unable to accept the feedback of an incorrect operation without being totally disrupted by it; for this reason, the protomental system attempts to split off the signal, even if it is in consequence itself liable to 'fall to pieces'. In more advanced affective terms, this would be equivalent to uncontainable rage. The attempt to silence the signal by evacuating it from the system is yet another incorrect operation, which confuses and at the same time amplifies the signal.

In other words, at cognitive level, the paranoid-schizoid circuits, which have the particular property of escalating persecution, represent the relative incapacity of the system to use error feedback for self-correction. The mind system has yet to learn how to receive, and indeed welcome, the experience of guilt, and to tolerate the resulting unpleasure for the purpose of using it – that is, of developing more appropriate operations. This learning process is the 'depressive situation'.

9.2 Proto-guilt and 'error'

Different shades of meaning are conveyed by the terms 'guilt', 'culpability' and the 'sense of guilt'. These must be distinguished if their use by the various schools of psychoanalysis at different times is to be clearly understood. As we have seen, they are used in different senses by early Freud, late Freud, Klein and Bion, as well as in our own description.

'Sense of guilt' emphasizes the subjective experience, whereas 'guilt' and, to an even greater extent, 'culpability' seem to apply to an objective fact – a transgression, mistake or error likely to have done harm. Hence the legal usage of these terms and the obvious fact that there may be a 'sense' of guilt even without actual guilt, and vice versa. In psychology, whichever

and possesses a modicum of explanatory value, is intended to show how the mind at this level contains 'something' complex and ineffable, and hence also indescribable – a monstrous internal object (a 'bizarre' object, in Bion's terms), a mental content basically constituted by processes of emptying out of the mind, which is therefore an antimental entity (–K, in Bion's notation). This is the nature of the proto-guilt arising in the mind.

term is used, we are necessarily referring to subjective events, and here the first problem arising is to identify the extent to which they are conscious rather than unconscious. Psychoanalysis, of course, considers them only in connection with unconscious experiences. Whereas this was relatively simple in 'early' psychoanalysis (the theory of repression), the problem has gradually become more complex with the ever more subtle correlation (or perhaps replacement), in later psychoanalytic developments, of the 'depth', so to speak, of the unconscious, or 'degree of unconsciousness', with the primitiveness of the processes organized in the course of psychic development, and with the progressively greater possibility of investigating and describing them. In this context, the experiences identified by Klein, which are unequivocally unconscious, are not considered according to their greater or lesser degree of unconsciousness, but from the point of view of their being much more primitive than the processes described by Freud. The Bion school places even more theoretical and clinical emphasis on identifying the deep processes on the basis of their primitiveness and of the fact that they constitute developmental functions for knowledge, both at primal level in the infant (diachrony of development) and in adult mental functioning (synchrony of deep thought processes), rather than destructuring or psychotic functions. Similarly, our description stresses the progressive construction of cognitive functions (the level of mental operations).

Here, however, we refer to 'incorrect' or 'erroneous' operations, having the objective effect of destroying the nascent mental structures, and to the possibility of the relevant signal of destruction being used for the purpose of not destroying rather than being expelled so that they continue their work of destruction. In other words, we seem to be referring to 'objective' guilt, resulting in harm, so that a possible distinction between guilt and the sense of guilt appears essential. However, the distinction called for in our description is different. In the case of such primordial processes, remote not only from any 'sense' or feeling at all approximating to a conscious sense or feeling, but also from mental states in which a self is distinguished from any 'object', the notion of a 'sense' of guilt refers to something whose qualification as an 'experience' is dubious, in so far as it cannot be observed as such, but only inferred – that is, inferred on the basis of what we know (or postulate) about the formation of mental processes and, in particular, by clinical psychoanalysis of the Bion type. In that situation, these 'somethings' take shape in the analyst's mind as if they were experiences – namely, affects and feelings. That is how he experiences them, and, in accordance with that experience, attempts to communicate them back to the patient in the form of interpretations. However, the idea that, as originally manifested in the patient, these 'somethings' were actually 'affects' can be asserted only a posteriori, once they have already been transformed by the analysis. These states are therefore remote from any possibility of being 'felt' by the subject, even with the aid of interpretations. This is all the more true in the case of a neonate than in that of an adult in analysis.

This use of the words 'sense of' thus proves to be improper. They could at most be applied to thoroughly stabilized reparative depressive mechanisms (see below), in which a mental rather than a protomental sense of guilt could be discerned. At protomental level, conversely, the 'guilt' involved is one that could become mental, but can readily be dementalized by paranoid-schizoid mechanisms (cf. Bion's evacuative processes, –K functions and beta screen).

A sense of guilt proper exists when a mental event arises in the form of a signal of something that has happened in the mind (and is perceived there), involving the infliction of harm on something experienced as good – that is, when the signal is recognized as such. From this point of view, the sense of guilt implies a motivation to 'remedy' the situation – that is, its depressive elaboration. In the case of more primitive processes – protomental or premental processes (or indeed antimental processes, such as Bion's –K) – it would be more correct to refer to antecedents of the sense of guilt, basal anxiety (perhaps the basal anxiety of the entire mind), proto-guilt or guilt-related anxiety. Grinberg (1976: 71f.) notes that in her late work, Klein

> admits the appearance of early guilt. This might appear to be in contradiction with what she maintained on so many other occasions – that the origin of guilt is based on the integration of the partial aspects of objects into a total object. In other words, on the one hand she affirms the need for an ego integrated and developed enough to be capable of experiencing guilt, and on the other she says that it is possible to feel guilt in a premature form at the beginning of life. I believe that this contradiction can be explained only if we admit the existence of the two types of guilt, a concept that is very close to Klein's, even though she does not mention it explicitly. [. . .] Depressive guilt, with all the characteristics described by Klein, requires an integrated ego in order to be fully experienced and used with its reparatory effects; whereas persecutory guilt shows itself early, with an ego that is still weak and immature, and increases automatically with the anxieties of the paranoid-schizoid phase or in the face of any frustration or failure in the development towards the depressive phase.

In the context of our theory, we shall therefore refer to guilt 'events', as well as to senses of guilt, although it must be borne in mind that this term can refer to processes very different from the ones thereby denoted in traditional psychoanalysis. This guilt will also be regarded as the actual destruction, or susceptibility to destruction, of the nascent mental structures. It can then legitimately be regarded as synonymous with the concept and term 'mistake' or 'error', because it constitutes a cognitively erroneous operation – that is, one that leads not to the knowledge of reality but to its falsification. On the other hand, in our view, the term 'sense' is appropriate for denoting the 'signal' – that is, the fact that the 'cognitive error' can be recorded and used to avoid further errors and destruction (that is, to remedy the guilt error), rather than going unrecognized and being expelled from the mind. In other words, it seems justifiable from a cognitivist point of view to take account of both levels – the subjective level, which is clearly expressed by the term 'guilt' or 'sense of guilt', and the objective level of the system's logical operations, which is well described by our use of the term 'error' or 'mistake'.

Let us now consider the formation of the above events at the various levels of protomental functioning, as well as the possible configurations they may assume on the basis of images drawn from the various types of bodily experience. One possible antecedent is that the state of hunger takes the form of a looming monster, which the neonatal mind casts out and attempts to keep at bay. In this case, the prototype of the signal that ought to signify that the baby is hungry is totally avoided, because at this first level the mind, instead of receiving it, evacuates it, seeking as it does not to feel anything unpleasurable. We may therefore invoke a primitive possible guilt anxiety at this first level.

A second situation is the state of 'rage' arising, as described above, when the neonate is engaged in the process of distinguishing between memory, hallucination, imagination and perception: he uses the memory trace to perform hallucinatory exorcism, but this cannot succeed because the hunger lasts too long. The neonate then experiences a state that tends to cause him to destructure the memory trace (to evacuate it), thereby depriving him, failing sufficient compensation for this mechanism, of the recently acquired capacity to recognize the nipple when placed in his mouth. In this case, the state of failure, which can be conceived as a precursor of a state of rage, ought to be a signal that can be decoded as guilt – that is, decoded in such a way as to produce an error signal in relation to the baby's notion that he has created the breast himself and indicating that the breast must instead be sought in the outside world. That will be the result if and when a stable capacity to distinguish between memory and perception develops. In this case, the signal will have been metabolized so as to recognize reality – that is, to construct the perceptual process. If the signal is not used, however, the mind may operate autotomically, destroying its own memory trace. Hence, a delay in the onset of the perceptual capacities, or their failure in adult psychotics, suggests the working of (hallucinatory) processes indicating that the prototype signal of guilt was not accepted.

A third possibility may be attained in the paranoid-schizoid position, in which the attack, or rather the counter-attack in response to persecution, is also projected; in other words, the mind is unable to realize that it is attacking something of its own, so that the persecutory, devouring breast is formed. The persecution gives rise to further counter-attacks, which are in turn projected and increase the level of persecution. The paranoid circuits are a typical example of a situation in which the signal that ought to supply information on what the mind is doing is used for a purpose other than that of providing information. The result of the 'error' is a progressive increase in the sense of persecution. Moreover, this is a telling example of how an 'error' that is not signalled leads to further errors, or, in other words, of how the concealment of the error – that is, of what could have signalled an incipient experience of guilt – leads to a repeated series of ever more serious errors. Again, the failure to use the signal of guilt for the purpose of attacking leads, error by error, to the misrecognition of reality. Reality then becomes persecutory: an individual in the throes of delusion is unable to take cognizance of the external world as it is, but will inexorably experience it as confused with his own bad internal phantasms.

We have thus described three possible forms of constitution of the 'antecedents of guilt'. The origins of these 'products' of the mind system, which will later give rise to the so-called sense of guilt, can be identified in more detail in the eight levels of protomental functioning distinguished in Chapters 5 and 6.

A particular form of processing of guilt signals is the infant mind's use of the very common experience of stomach-ache. It may be postulated that this bodily experience inherently tends to take the form of an internal experience, even though the mind may seek to evacuate it in schizoid fashion or to project it along paranoid lines. After all, the painful afferences on the basis of which the experience can be constructed are too intense to be readily exorcized or destructured, or in some other way deprived of their meaning as an unpleasurable experience internal to the body. This meaning is, so to speak, rooted in biology, unlike the situation with other perceptions, which are constructed on the basis of afferences that lack a neurological predisposition to convey meaning. Stomach-ache thus possesses its own 'natural' tendency to constitute an unpleasurable internal perception. At its primitive levels of functioning, the mind will obviously be led to experience stomach-ache as a monster that has arisen inside it or penetrated into it; in this experience, the visceral afference comes to form part of the same constellation of oral afferences that constitute the first internal object. This may therefore be the monster that convulses the baby and hurts him inside (in his stomach). To counter the primitive bad breast, the mind will attempt its customary exorcisms, but these are less likely to succeed when the monster entering into it consists of afferences not only of hunger but also of stomach-ache. The stomach-ache afferences therefore tend to compel the mind to see them as an experience belonging to the inside of the child's own body and hence to the self.

Stomach-ache could then constitute the foundation of an unpleasurable experience felt to be internal and not evacuable by exorcistic operations, whose efficacy is therefore invalidated. It thus becomes the prototype 'signal' of the inappropriateness of such mental operations – that is, a signal of 'error' – and consequently opens up new perspectives, in the form of mental operations more appropriate to reality. On the other hand, the proto-rage due to the chagrin[4] resulting from failed exorcistic and hallucinatory operations, and hence also the first attack on the object (either external or its internal representation), becomes bound to these unpleasurable experiences. An infant suffering from stomach-ache is also an infant enraged at having stomach-ache, who attacks and combats the badness he thinks has got inside him by means of this 'proto-rage', or rather by means of the incorrect operations that constitute it. This 'rage' becomes bound to the badness, and if the badness cannot be cast out, the signal constituted by this event contains within itself the meaning that the rage is ineffective, and that it is therefore an 'error'. In so far as the signal of unpleasurableness is accepted, the rage can be transmuted into the experience of something mistaken, which serves no useful purpose; indeed, it increases the pain[5] because it results in the disappearance of whatever good might still be in the mind – namely, a good memory trace that can be used to endure bad experiences. The error then becomes an experience of guilt and destructiveness.

This meaning becomes clearer when the mind moves on from purely schizoid to paranoid-schizoid mechanisms. At this level, the attack on the object is already much more clearly delineated: oral fantasies, later also visual, prehensorial and masticatory fantasies (see Chapter 8), and later still anal and other fantasies, are metabolized in the paranoid circuits in order to get rid of unpleasurable experiences, to the detriment of the perception of reality – that is, of knowledge. An experience such as stomach-ache, with its 'internal' intensity, thus tends to block the paranoid-schizoid mechanisms and to act in the opposite direction. It might therefore facilitate the binding of destructive (envious and greedy?) fantasies to an experience of internal badness. These fantasies would in this case take on the significance of 'badness' – in other words, they could be recognized as what they are in reality, as destructive fantasies. Furthermore, their action as a projective counter-attack on the persecutory breast would be shown to be inappropriate; in other words, their 'erroneous' character would have been signalled. Such fantasies could thus assume the significance of something bad belonging to

4 As stated in Chapter 5, this is an adultistic term for something that is actually much more primitive.

5 The significance of rage as 'erroneous' and not 'serving any useful purpose' should not be understood in an adultistic sense. On the other hand, this meaning – error – is attained within the system after metabolization in paranoid-schizoid circuits. In other words, it may be projected, thus confirming the existence of an 'enraged' persecutory breast that is even more tremendous and must be cast out and fought against. Such 'rage', in this case, increases the level of pain, which must be evacuated; it cannot then be accepted and depotentiated by recognition of the error.

the self – that is, they could clothe themselves in an experience of guilt, which, in turn, could constitute the basis for them to be recognized as an attack on the very objects that are also desired, which thus become 'known' as the same objects as the ones that are hated. That would put an end to the split, within the self, between good and bad fantasies and between images of good and bad objects, and, outside the self, to the artificial split whereby the object is divided into two opposing objects; in this unification, the object can now be known as it objectively is. Moreover, it will then be possible to recognize the destructive fantasies as destructive of the very internal objects that serve to represent reality appropriately – that is, of the cognitive structures. In this way, the infant moves on from paranoid-schizoid to depressive dynamics.

The prerequisite for the entire process described above is that the experience of badness shall have been recognized as a signal of error and destructiveness – that is, of guilt. The experience of stomach-ache, like other experiences, can then be seen as an element that interferes with the paranoid-schizoid metabolism, becoming established as a signal of a wider system of signals, such as those gradually arising as and when experiences of guilt become more complex and ramified in the depressive position. The experience of stomach-ache can therefore be regarded as a bodily antecedent of the model of depressive processing of the signal of guilt. However, this possibility of using a painful experience positively has limits, represented on the one hand by the intensity of the pain and on the other hand by the complementary capacity of the system at the relevant time to use it.[6] In other words, the thesis is not that a child who frequently suffers from stomach-ache will develop better, but simply that the experience of stomach-ache, which occurs in all children, can be used by the mind in the manner described above. However, if the painful experience exceeds a certain threshold,[7] relative to the system's capacity to tolerate it, the result will be, not a further step toward a depressive situation, but instead recourse to even greater persecution and more desperate autotomy. Moreover, the system's capacity is correlated with the context of communication – that is, with the function of making bad objects good that is performed by the caregiver's reverie (see Sections 10.3 and 10.4).

6 It would be interesting to consider in depth the hypothesis that stomach-ache is not merely a bodily event that is then used for the mind's operational purposes, but actually constitutes the somatic expression of certain mental operations.

7 The intensity of the pain may exceed this threshold if the child not only suffers from 'ordinary' stomach-ache but is also affected by other painful experiences, such as physical injuries, burns or other pathological states. It is therefore important to bear in mind the mental risks to which such infants are exposed, as well as the psychological care they require as contrasted with the organization of our infant health services.

9.3 Processing of guilt and bodily experiences

The antecedents of the sense of guilt, or 'signals of guilt', are described above from two standpoints. The first concerns the bodily experiences whose afferences underlie the processing of such signals, with particular reference to bodily situations which, by virtue of their specific anatomy and physiology, are most suitable for the constitution of prototypical experiences of guilt (Chapter 8: eating, swallowing, looking, taking, breaking, nibbling or stomach-ache). The second has to do with the form of metabolization of these signals by the incipient mental structures (Chapter 6). The mechanisms whereby the mind 'has experience' depend on the various bodily events on the basis of which the mind, from one occasion to the next, becomes able to have experience on successive levels. In other words, the relevant bodily zone and its function determine the construction of the different, progressive mental functions, which in turn determine the possibility that certain bodily zones and functions can be used for the further construction of experiences and mental functions. One and the same bodily experience may be metabolized on these different levels, as we have attempted to show by our description of the experience of stomach-ache.

All the experiences of the most diverse bodily functions can be processed on the various levels, and some of these functions tend to move from one level to another: not only is one and the same experience processed at various stages, but, depending on the specific experience, the mind tends to function on one level rather than another. We must therefore consider which experiences may favour the further development of the mental structures – that is, induce the mind to move on to new levels of functioning. Neonatal stomach-ache is described above in these terms. Although, admittedly, any experience can be metabolized on the most primitive levels for particular reasons, it is surely worthwhile to try to identify the experiences that favour the development of more advanced mental operations.

In addition to those already mentioned, these include an experience that an infant can easily have – namely, that of the empty bottle, especially if a state of hunger or a need to suck (which is more easily satisfied by the nipple) lasts for any length of time. The empty bottle is the focus of an immediately bodily experience involving the mouth, sucking and hunger, and at the same time of a relationship with the body (the overall tactile experience) that is more distanced than in the case of breast-feeding. In artificial feeding, the more stable the visual perception of the bottle becomes, the more easily the baby will come to see it as an entity different from himself. Visual perception also facilitates discrimination, when the baby begins to distinguish the full from the empty bottle. This experience involves the possibility of linking the need to a wish, and the wish to the use and consumption of the object; in this way, a meaning is assigned to the subject's action. The mind can experience as its own something that actually is a function of its own and can link the ensuing fantasies – which may be greedy and perhaps also envious – to this experience; the infant will thus be able to experience a series of unpleasurable events as his own, as products of the self. In the perception of the empty bottle that may arise in this context, the infant also has an opportunity to have the experience of 'having consumed' the good that has been used; on the other hand, it now also becomes possible for him to see his own fantasies as modulated by a non-consuming relationship between himself and a differentiated, external object.

Such an experience of using the external object may, in other words, be assumed not only to serve the purpose of consolidating the perception of the external object as such, but also to act as a parameter for the modalities with which to establish a relationship with it, involving a use that is not omnipotent, let alone destructive. This means that the greedy and envious fantasies respectively are experienced 'negatively'. In this connection, the experience of finishing the bottle may be treated by the mind as a signal informing it of the quality of the operations performed and of the preservation of the object. The signal will then be processed on various levels, but it can readily be imagined that the experience in itself may contribute to a shift toward more advanced levels of mental operations.[8]

Another, perhaps more telling, experience is when the infant breaks objects after having tried to grasp them. On the level of processes more advanced than the metabolism of prior experiences permits, breaking an object lays the foundations of an experience of something erroneous that has taken place in the mind and in the action stemming from the mind, which involves the destruction of the relevant object. In other words, this experience can constitute a sufficiently solid foundation for an experience of guilt and the unequivocal signal of the possibility of destroying. This idea may be compared with the common observation that a child breaks objects because he is angry with them, because they do not obey his wishes or his attempts to take them in his hands, and that the breakage is often due to a clumsier and manifestly angrier movement than those he has learned to perform. Having broken the object, the infant is dismayed: does this constitute guilt? Often, however, after breaking the object, the infant is even more furious than before – owing to projection, persecution and counter-attack.

Many other experiences, too, can be regarded as occupying a privileged position in the constitution of internal signals of incorrect operations. A detailed discussion could be devoted to these experiences alone; it would

8 Breast-feeding of course presents other advantages over artificial feeding – mainly, perhaps, that it encourages mother–baby interaction and hence the metabolization of communication (making good objects bad, reverie, projective identifications, etc.) through various sensory channels, in such a way that the infant learns a language whereby he can confirm the acquisition of a reality different from himself.

include a systematic description of the entire process of symbolization with its progressive expansion not only of modes and levels but also of contents – that is, fantasies, images and engrams – which the progressively acquired functions allow the infant to assemble and construct, and whereby the functions themselves are further expanded and ramified. To use an engineering metaphor, advanced technologies here yield increasingly sophisticated materials, which in turn stimulate the invention of further construction technologies. Besides the bodily zones outlined above, particular importance will attach to sphincter-related experiences, which have been somewhat vaguely subsumed by psychoanalysis within the so-called anal phase. These vicissitudes will be very important for the mind's processing of its own aggressive experiences. I have attempted elsewhere (Imbasciati 1983b) to describe in more detail the use of the various bodily functions for symbolopoiesis.

To conclude this part of our argument, it is emphasized that the situations analysed here entail the possibility, on successively more advanced levels, of 'understanding' a mistake and 'feeling' guilty about possible destructive fantasies or actions. However, in order for such an experience to become established as such – that is, as a mental experience – it must escape falling victim to autotomic operations, or, at least, the mind must not perform total autotomy. In other words, the system must accept as its own the tendency to get rid of such signals and must therefore abstain from getting rid of them; it must consequently retain within itself the pain thereby entailed, at the same time acknowledging – that is, containing – rage and aggression. According to Harris and Meltzer (1980: 390f.), if the functioning of personality is considered in dynamic terms, the

> processes of development which *bind* mental pain in the form of structure are further amplified by mental mechanisms aimed variously at modulating, modifying or evading mental pain. The *modulation* is accomplished primarily by thought leading to understanding and actions that may successfully modify or adapt to the external world or internalize new qualities into the internal object that comfort or strengthen the personality. *Modification* of anxieties is achieved through *omnipotent phantasy* of various sorts called the *mechanisms of defence*, but many of these same defences may be employed in a more violent way to *evade* mental pain through its distribution either in the internal world or in the external.

The processes of modification and evasion of pain are of course remote from the consciousness that may characterize the more mature sense of guilt in an adult, even if they are a prerequisite for it. The phenomena described above are absolutely non-conscious processes, which, moreover, are the only kind that can occur in a neonate; the words 'understanding' and 'feeling' are to be understood in this sense. They refer to experiences involving a recognition by the mind system of something that belongs to itself.

As already stated several times, it is utterly inappropriate to invoke consciousness at neonatal level; conversely, it is meaningful to consider how 'experiences' may be had as such in relation to a consistent internal entity that we call the self, or alternatively be expelled from the mind. In order for certain internal events to become constituted as signals of guilt, the mind must accept an unpleasurable emotion, and thus interrupt the form of functioning directed toward the avoidance of unpleasure and bearing the stamp of the paranoid-schizoid mechanisms. Unless autotomic operations are interrupted, there is no true experience of guilt. The moments at which destructive events take place may be mental events that are used and processable only if they are not split off and expelled; otherwise, such events remain or become non-mental, or rather anti-mental. In other words, events occurring in the mind may undergo two possible vicissitudes: they may be either transformed into thoughts, or expelled[9] and deprived of the possibility of becoming mental. Bion (1962, 1963) used the corresponding notation –K for processes occurring in the context of an anti-knowledge link, as a part of the negative 'grid', whereby each category of the 'elements' of thought can be transformed into a 'lie'. The same notion is inherent in the hypothesis, advanced by Harris and Meltzer (ibid.: 390), that it

> is possible that along with the structuralizing of the personality [. . .], a parallel development takes place, which evolves an alternate 'world' on the basis of negativism and envy, the delusional system of the schizophrenic part of the personality. Its prominence in the whole structure, its access to consciousness, and its indirect influences on development vary, of course, from individual to individual from time to time [. . .], but they will be assumed [. . .] to be ubiquitous.

The two systems operate simultaneously and in parallel on the same experiential data, but draw different conclusions from them: on the one level, thought is processed by learning from experience within the world of emotions, meaning and object relations; while, on the other, all the data are forced into, and interpreted in the light of, the basic assumption of the subject's own omniscience, so that nothing new is learned and the delusional system of the lie proliferates like a parasitic mind within the host mind.

Paranoid guilt is thus paradoxically definable as a type of guilt without the

9 The process of evacuation described by Bion and Meltzer can be compared to our concept of autotomy: in information-processing terms, it entails the deletion of something that had been, or was about to be, processed. As with a computer virus, all memory, even of important programs, is lost. The neuropsychophysiological trace is probably altered by a corresponding protein metabolism that modifies RNA synthesis and resynthesis, or perhaps the neuromodulation of synaptic circuits.

mental experience of itself. Guilt that is more real will be experienced when the mind succeeds in stabilizing within itself an object in its wholeness and uniqueness; this will signify recognition of the subject's own sense of guilt as distinct from the object. An infant, in the paranoid-schizoid position, makes no distinction between an object and the fact of – his own and not the object's – experiencing of an unpleasurable sensation. The capacity to distinguish the subject's own mental operation from the object, and from the image of the object to which it refers, thus becomes a criterion for describing the experience of guilt, which proves to be a fundamental element in the recognition of reality, affording access to the world of object relations typical of the depressive position.

9.4 Acquisition of the signifier by splitting-cum-idealization and manic denial

With the achievement of the distinction between self and not-self (level 4 in Section 5.3) and the subsequent construction of the perceptual capacities allowing the not-self to be constituted as external reality (levels 5 to 8), the internal objects are no longer experienced in a confused jumble as if each individually included the entire universe, but are differentiated and placed in relation to something precise and delimited, which is perceived as existing outside the subject himself. In initial mental aggregates, it is now possible to distinguish two categories of aggregates linked by the same type of relationship as that between the signifier and the signified in Saussurian linguistics: de Saussure conceptualizes the linguistic sign as an entity comprising the signifier – the 'acoustic image' – and the signified, or 'concept', which corresponds, in the extralinguistic sphere, to a referent (such as a book as an object, as opposed to the concept of a 'book'). In the development described here, the internal object takes the form of a signifier in relation to the external object, which can therefore assume the status of a signified. The external object is then important not so much in itself, as a referent, as for the manner in which it is known – or signified – in accordance with the organization of personal experience that has come into being internally as a signifier, which, precisely, is capable of signifying, or attributing meaning to, the external object. The first mental aggregate, or the first internal object(s), is a signifier that is mixed up with the signified, since there is no distinction between the self, in which the signifiers are aggregated, and the external world from whose referents the signifieds are drawn. On the most primitive levels, the first engrams laid down in the mind system cannot yet be used for distinguishing between signifier, referent and signified; they are therefore treated as if they were equivalent to referents, so that there cannot be any real 'knowledge'. In order for differentiation to come about – that is, for the mind to treat signifiers in relation to signifieds from the external world – what is formed in the mind must be retained within it, but not treated as if it were a referent to be cast out and expelled if unpleasurable. The internal objects serve for knowledge in so far as they become differentiated into signifiers relative to signifieds; on a primitive level, the latter are much more like referents – that is, concrete objects that are retained or destroyed in accordance with the unpleasure principle, as if they were the actual food that enters the mouth and that can be expelled. In order for knowledge to be possible, the self must be distinguished from the external world; this distinction consists in the possibility of retaining in the mind some kind of trace of all experiences, including painful experiences, so that they can become signifiers of signifieds that can thus be known. This knowledge places them in relation to referents, but what is important is the signified and not the referent. A particular vicissitude is undergone by signifieds that lack an external referent – that is, ones referring to internal products such as affects, fantasies, experiences or 'thoughts'. These signifieds constitute the basis for the organization of what will be illustrated below by the introduction of the concept of intrapsychic permeability (see Section 9.5).

The first painful experiences assume the form of bad objects and are expelled by the process of paranoid-schizoid metabolization, with consequent impoverishment of the mental processes. The capacity to retain such painful experiences in the mind and to process them mentally is achieved by successive cycles of splitting and idealization (see Section 7.4), and subsequently by manic denial. The first step in the transformation of the bad breast into a mental content comprises the non-evacuation of this unpleasurable experience, which at first is not distinguished from the external object to which it is referred. Only adults can distinguish the image of a thing from the emotional reaction to it; infants, on the other hand, confuse the image of the bottle with the rage caused by its failure to arrive. If the bad internal object is no longer expelled, but instead treated as something belonging to the self, like the good object, the infant is then confronted, within himself, with a good bottle, the memory of which he easily preserves, and with another – angry – bottle which is capable of attacking and destroying and is therefore in turn attacked and eliminated. The latter 'breast' was at first immediately evacuated, but later, certain experiences – in particular, those described in the previous section – present it again to the mind system, reintroducing it in such a way that the system is in a position to contain it. If the mental operations develop positively, the mind makes use of the opportunities afforded to it by certain experiences to embark on operations directed toward the containment of bad as well as good objects within itself. In so far as this function succeeds, the mind acquires the capacity to have signifiers of the 'bad' – that is, to recognize its own unpleasurable states.[10]

10 The quality denoted by the adultistic adjective 'unpleasurable' is of course subsequent to a process-related attribute that could more accurately be described by the word 'harmful'.

A primitive mode of containing bad objects within the self comes into being because the bad object must coexist with the good one: the subject fears that the bad object, which gradually comes to be retained in the mind and is experienced as destructive (oral, devouring), may attack and destroy (devour) the good objects. For this reason, the subject can begin to contain it only on condition that he keeps it drastically separate from the good object inside himself. The mind, as yet unaccustomed to treat the bad object as its own and as a signifier rather than a referent, in effect still sees it as a monster, which it no longer attempts to exorcize, but retains subject to the taking of appropriate precautions, keeping it well away from its dear ones, as befits a monster to be accommodated in one's own home.

Some of the above descriptions are presented in terms of fantasies and feelings, with anthropomorphic metaphors in accordance with psychoanalytic tradition. This approach will not be pursued much further. However, a more strictly cognitive interpretative hypothesis is also possible, in which the description given in Section 7.4 is translated into terms of cybernetic logic.

The mind has tried and tested a selective system of expulsion of the bad and retention of the good: any attempt to retain the bad presents a threat to the system. The bad does not cease, but persists, superimposed on other manic operations, as we shall see below. Confusion is therefore likely in the selective action of the mechanism of expulsion, with a consequent probability of expelling the good; the mechanism of clearly separating good from bad could thus be interpreted as a temporary operation directed toward reducing this risk of confusion. Such a risk may also be inherent in possible attempts by the mind system to put good and bad objects together, with a view to achieving the end result of the process of depressive restructuring – namely, the reunification of good and bad into unequivocal signifiers of an object that actually exists in reality and is appropriately signified (known) to the mind independently of the mind's reactions to it.

Whatever the logical explanation of these phenomena, there is the seeming paradox that, precisely in an intermediate phase of the reunification of good and bad that is intrinsic to the mind's depressive restructuring, the mental aggregates to be reunified in fact exhibit a greater degree of separation.

In the logical notation used in earlier chapters (see Section 7.4), the operation represented by:

$$B_{i_g}\Big)\Big(B_{i_b}$$

is followed by preservation of the distinction between B_{i_g} and B_{i_b} – namely:

$$B_{i_g} \neq B_{i_b}$$

Hence also:

$$B_{e_g} \neq B_{e_b}$$

The two classes of objects, or worlds – internal and external, good and bad – are thus kept not only at a distance, but also as separate as possible. The purpose of this operation is to preserve the good, even if the resulting separation proves to be excessive and entails operational deficiencies.

The function of this split is to avoid any contact between bad and good experiences, lest the former destroy the latter, giving rise to confusion and attacking thought and the memory of the good object. This, according to Bion (Bion, F. 1984), is paramount, constituting as it does the foundation of development: the establishment of a process of splitting-cum-idealization, whereby good and bad can be distinguished (albeit in exaggerated fashion), makes it possible to overcome the first level of confusional problems typical of the developmental sequence. As Segal (1964b: 35) points out:

> It is splitting which allows the ego to emerge out of chaos and to order its experiences. This ordering of experience which occurs with the process of splitting into a good and bad object, however excessive and extreme it may be to begin with, nevertheless orders the universe of the child's emotional and sensory impressions and is a pre-condition of later integration. It is the basis of what is later to become the faculty of discrimination, the origin of which is the early differentiation between good and bad.

This experience of separation is analogous to that of a severely psychotic adult who feels divided into two persons. Since the neonatal mind is made up solely of these contents, a split between them has the consequence that the entire person – the neonate's self – is divided into two. The good part, the first thought structure, must therefore be enabled to continue to exist without becoming mixed up with the bad; in this way, it becomes possible to tolerate the bad object, because there is also the good one. It is as if the infant were thinking: 'I am bad, but I am also good' – bearing in mind that this is a contraposition and not a union, in which the emphasis is placed not so much on the element 'I am' as a recognition of personal integration, as on the alternation of good and bad states. A trace of this division into two can be discerned when a child is heard to talk about himself as a good child and a bad child, as well as in dreams, including those of adults, in which the dreamer feels that something is at one and the same time its opposite.

The separation of good and bad also extends to the internal objects constituted by mental operations themselves. The mind must therefore keep thoughts that 'work' separate inside itself from ones that 'do not work', because the former are the creative thoughts that give rise to new mental operations (symbolopoiesis) and develop the mind's capacity, while the latter are destructive, triggering the paranoid-schizoid circuit.

Again, whereas, from the point of view of logic, retaining the bad object inside the self gives rise to confusion, in affective terms it generates something resembling fear and terror, because these 'thoughts' are expressed in bodily fantasies experienced as real facts – that is, as tearing another body to pieces. This other body is readily referable to the other body known to the infant – namely, the mother's. Moreover, the destroyed objects are a source of endless persecution, so that the infant finds himself living in a world of dead objects, with the terror of being left alone – and indeed, of being left alone with something that is, precisely, terror (Klein 1963). Meltzer (1973: 148) defines this type of psychic suffering as

> a paranoid anxiety whose essential quality, paralysis, leaves no avenue of action. The object of terror, being in unconscious phantasy *dead* objects, cannot even be fled from with success. But in psychic reality the vitality of an object, of which it may be robbed, can also be returned to it, as the soul to the body [. . .].

Now just as the experience of the good internal breast entails the existence of a good external breast, it may be thought that the same applies to the bad internal breast. However, the bad external object is no longer the non-mental primitive persecutory object. The mental configuration is now different, and presupposes the representation of a bad referent presumed to exist in objective reality, which the mind attempts to know or supposes to be knowable. Hence, the external object is split into two objects that are believed to be real.

All the external objects are now found to be divided and separated into two classes – of good and bad objects respectively. There will therefore be a good and a bad feeding bottle, a good and a bad hand, and a good and a bad mother. Even memories, of whatever object, will be treated as if they stemmed from different objects in the external world – that is, from different referents.

Such a state of affairs has already been represented by the following notation:

$$B_{e_g} \neq B_{e_b}$$

To illustrate more clearly the division into two, the split and the distance, the following notation may also be used:

$$\left. B_{e_g} \right) \left(B_{e_b} \right.$$

The splitting of the external objects absolutizes the qualities that determine which of the two fundamental classes an object belongs to, so that an object appears either totally good or totally bad. If the positive qualities are taken to extremes, this is said to constitute the mechanism of idealization. The split is thus reflected in the internal world by a corresponding division of the external world, with idealization of what is experienced as good. This is confirmed, for example, by the trace of this phenomenon discernible in adolescent disappointment with previously idealized parents: the former idealization, manifested during latency in a rich fantasy life, originates in very early infancy, in the idealization of the good object as opposed to the bad one.

Objects are split into the two classes of good and bad in order to defend against the fear and confusion stemming from the bad internal objects. However, this artificial division of the world does not overcome the contrast, or indeed opposition, between the classes of objects. The mind must therefore undertake a new mental operation, as a further refinement to that of splitting, with a view to reducing the opposition of good and bad and at the same time protecting itself from the risk of mutual destructive contamination of both. This operation takes the form of *idealization* of the good objects and *manic denial* of the bad.

This can be expressed in mathematical-type notation as follows,

$$(B_{i_g})^{n}\Bigg)\Bigg(\,(B_{i_b})^{-n} \quad \text{and} \quad (B_{e_g})^{n}\Bigg)\Bigg(\,(B_{e_b})^{-n}$$

in which the second and third of the four symbols are the most important: manic denial is applied essentially to the subject's own internal bad aspects, while idealization is directed toward external objects.

Idealization can be represented as a kind of realistic exaltation of good objects; this means that the bad objects are not kept separate, but that their intensity for mental life is diminished. In other words, they tend to be 'denied'. This does not mean that their status as mental objects is abolished, for these are not paranoid-schizoid type evacuative processes which expel mental contents and structures from the mind. In the process of denial, the mental event, whether it be a content or an operation, remains in the mind, but is in effect confined within a sealed compartment, as if it were sequestered in a dark corner or, in terms of operational logic, isolated from the operational set.

The denial is obvious in the case of all the bad internal representations: the bad internal objects can thus remain confined in the mind and their importance disavowed. Once again, these processes are not conscious, but involve a more primitive relationship of 'proximity' of the self to its own bad experiences. In this way, the mind denies to itself the importance and force of some of its own destructive impulses, fantasies and bad mental operations. This permits a kind of exaltation of all the good aspects of the self, over and above their actual vitality – a situation describable as mania and manic denial (Klein 1940). The term derives from certain adult psychotic syndromes in which the subject sees nothing bad or unpleasurable either in himself or in the world, and lives in a state of utterly carefree euphoria. Mania always involves a deficiency in the perception, evaluation and knowledge of reality.

However, the error inherent in a manic operation is less pronounced than in paranoid-schizoid operations, in so far as what is denied preserves in a way its character as a mental event, even if it is placed at as great a distance as possible from the self. The signal of guilt, too, like other mental events, is denied in the manic situation: the feeling is encapsulated in the mind and preserved in inactive form – although it is always liable to resume its true significance, as in the sometimes clamorous swings from mania to depression observed in certain adult psychotics. Splitting-cum-idealization and manic denial may therefore be seen as processes essential to an initial distinction between signifier, signified and referent – that is, between something that represents an initial experience of something mental, on the one hand, and what this experience signifies in relation to an external reality, on the other hand. However, although this distinction involves unrealistic signifieds and referents (because they are split), it puts an end to the 'substantiality' of the first mental objects, and therefore reduces the need for them to be expelled.

A mental equilibrium involving a degree of persistence of manic operations may continue into adulthood and, if not too serious, need not give rise to manifest cognitive deficits, for manic mechanisms obviously permit the development of the good internal objects represented by valid mental operations, which lead to perception and the use of sense data to form engrams that match the external world appropriately. This world can thus be signified in the mind, and therefore known in operationally valid terms. However, in manic – or rather hypomanic – situations (or characters), there may be a deficit, mostly indirect, in the capacity to assess reality. This is particularly obvious when a manic personality is confronted with the knowledge of animate objects, as in interpersonal relations. In this case, by contrast with the superficial facility resulting from the extroversion, the joviality and heedlessness that often accompany the manic character, intimate contact and, in particular, the capacity for intimate understanding of other people's personalities present appreciable difficulties. This deficit conceals another, more conspicuous one that is always exhibited by the manic personality – namely, the subject's sometimes far-reaching inability to experience and understand his own self. For just as the manic individual stubbornly denies his own sense of guilt, he almost always also denies parts of his own personality that may be quite extensive; these remain not merely unconscious, but split off and enormously distanced from the self. Signifiers too, even if acquired as such, are split off. The cognitive capacity of the manic personality thus appears reduced, at least in terms of its inward-directed aspects, as opposed to its orientation to the outside world: manic tendencies are frequently accompanied by deficiencies of insight and self-knowledge.

Mania may be seen as contrasting with depression, on a level different from that characterizing the opposition between the depressive and the paranoid-schizoid position. According to Segal (1964b: 83), mania will be used in defence against any experience of having an internal world or of containing in it any valued objects, and against any aspect of the relation between the self and the object which threatens to contain dependence, ambivalence and guilt.

The persistence of paranoid-schizoid mechanisms signifies that the cognitive mechanisms, and in particular the capacity for self-knowledge, have undergone substantial distortion; however, in the analytic treatment of such subjects, a keen interest in knowing is found to have been preserved. By contrast, this interest appears reduced in manic characters; indeed, it is blocked by an obstinate pseudo-rationality that prevents access to the internal world.

9.5 Reparation: repairing the engrams of thought

In our account so far of the development of the mind, we have described a succession of levels starting from a primitive internal object (the breast), which we have shown to break down initially into good and bad aspects and later into an internal and an external aspect. We have also shown that paranoid-schizoid mechanisms and the persecutory circuit are the mind's first mode of action, which subsequently gives way to the transitional phase of splitting and idealization of the internal objects and manic denial of the bad internal object. As we have seen, the result of these operations is the division of the breast, or rather of all internal objects[11] into two: hence, not only is an object split, but also, probably, all 'memories', or memory traces and all other mental aggregates and products are divided into two and kept separate, as if they did not stem from the same real experience.

This division of the internal objects into two inevitably implies the performance of a similar operation on all signifieds. The division into two of the signifiers – that is, the separation of the good from the bad internal objects – that is carried out to ensure that the 'bad' mental operations do not destroy or somehow damage the good ones entails a corresponding division of the signifieds: the external objects are known – that is, signified – as if they were divided into two, good and bad. The result is an alteration of the sense of reality, in that the mind lives as if in a world in which the referents too are so divided. The construction of the perceptual capacities admittedly favours the formation of unitary referents, but their construction is slow and laborious

11 Although in this book we sometimes refers to the 'internal object' in the singular, this is strictly speaking legitimate only with respect to the very beginnings of the construction of the mind, and perhaps only to the fetal stage. By the use of the singular, we thus mean the internal objects whose power can be glimpsed from time to time. Again, as already emphasized, the elements with which the mind works are in fact enormously complex, and their complexity increases progressively (see Section 8.5).

and, moreover, concerns the present object much more than 'memories'. In memories – that is, the traces making up the deep affective structure (that is, basic cognitive schemata) – the traces of split-off objects persist. The child's internal world is made up of 'part-objects' – in the sense not only that the wholeness and reality of the object have not yet been achieved, but also that the object is split into two parts in the processing of the information circulating in the system, as if one and the same external object gave rise to two qualitatively different types of knowledge. What Klein described as a part object because it is split in two by the affects – of hate and love respectively – is expressed, in our conception, primarily in representational terms (engrams) and in cognitive operations.

Klein followed Freud in regarding the affects as derivatives of instinctual energy, which was in turn attributed to biology; the affects were therefore seen as the motive forces that interfered with the mechanisms of memory and perception, which were believed to be natural and automatic, as they appear to a normal adult (naive realism), rather than constructed. For these reasons – that is, because she remained faithful to Freud's theoretical schema (and owing to her unfamiliarity with the contemporary findings of other psychological sciences) – she asserted that the affects gave rise to splitting, so that the cognitive aspect was also split off, and that this was so both in children and in adults affected by pathology. The converse is considered to be the case here. The affect is not the cause, but the effect, described in subjective-adultistic terms, of the splitting of mental operations that are kept separate, and the same presumably applies to the respective neurological traces. Klein's view that the child 'splits off' objects by a mechanism deemed pathomorphic is therefore incorrect: objects are in fact split in the neonatal mind because it has not yet learned to put them together (Imbasciati 1991: Section 3.2).

The failure to master reality by the mechanism of splitting and idealization may give rise to a situation of critical equilibrium, leading either to a return to the autotomic mechanism typical of the former paranoid-schizoid situation, or to a complete restructuring process involving a transition to a higher level of mental functioning, in which the split objects are reassembled and put together in a new and meaningful whole – that is, are repaired. Reparation then takes the form not only of an affect which repairs another affect that has proved to be bad (described as involving gratitude or love, as against hate and aggression), but also, in cognitive terms, of the mending, or repairing, of a functional set signalled as ineffective or broken, damaged by incorrect operations. The term 'reparation' is thus stripped of its almost moral connotation and takes on the precise and more scientific meaning of the repair of something broken. The mind-computer has the capacity to repair itself.

The fact of describing reparation in these terms does not mean that it can no longer be described in the subjective terms of an affective dynamic. The problem here is to avoid considering this subjectivity in adultistic terms. The reunification of good and bad objects gives rise to the experience of something that can readily be described as pain or guilt at the destruction wrought by the previous mode of functioning, which was directed toward the avoidance of suffering and frustration.[12] It is by no means easy to move on from flight from pain to the use of a mechanism that gives rise to suffering because the good external breast is equated with the bad internal breast. In other words, in this situation the child (in subjective terms of affects), or the system (expressed in logical-cognitive terms), is compelled to observe that the breast that provides so much good (or the engram corresponding to it) is the same as the breast which, when absent from perceivable reality, seems to inflict so much pain (that is to say, the engram of the good breast is compared with the signifier of hating it). This is a difficult transition, necessarily involving suffering, in which the infant must feel that the bad internal object is a function of his own that destroys both good objects and efficacious engrams.[13] The memory traces whose meaning assures the infant that there is no suffering, or that the suffering need not be feared (the experience of waiting) must now be put together in his mind with others that *do* symbolize suffering, whether experienced as the disappearance and deprivation of the good or, at a more mature level, felt, or about to be felt, as destruction – not only outside but also, and in particular, inside, where it has internal effects on the mental structures themselves. In 'mathematical' terms, the need is for a transition from the situation of splitting

12 Analysing the development of Klein's theory in relation to the concept of 'reparation', Meltzer (1979a: 129, translated) points out that it is connected with that of the concept of 'position': 'Whereas the first use of the "position" concept was of a developmental character, connected mainly with the concept of defence, and the second use was bound up essentially with psychopathology and the concept of fixation, the changes in the use of the term "position" in the 1940s and subsequent years place this concept in the economic sphere'. In parallel with this, Klein moves on from the concept of 'restitution', which means giving back something that has been stolen and putting back what has been taken away, to that of 'reparation', denoting the repair of damage in order to avoid persecutory anxiety. A final transformation is accomplished in 'Envy and Gratitude' (1957) and *Narrative of a Child Analysis* (1961), in which 'reparation began to take on a more mysterious meaning, something that happens in the depressive position. That frame of mind of depressive feeling, guilt or remorse or just regret, or wishing it had not happened, seems to make possible the process by which objects repair one another. The distinction between the active and passive components immediately galvanizes the understanding of the difference between manic reparation and true reparation. The true reparation is something that happens when the mental condition, the mental atmosphere is conducive to the objects repairing one another' (Meltzer 1978: 47).

13 Klein presupposes that the affects, presumed out of homage to the Freudian theory to be an emanation of the instincts, give rise to splitting, resulting also in a consequent split-off cognitive aspect. In our view, affect is an adultistic, subjective-affective account of what our logic-based, information-processing approach describes in terms of the destruction, separation and annihilation of mental operations. The affect is their effect and not their cause.

$$B_{i_g}\Big)\Big(B_{i_b}$$

to one that can be denoted by

$$B_{i_g} \equiv B_{i_b}$$

in which the good internal breast is equated (as indicated by the sign of identity) with its bad counterpart. This means that the mind is putting good memory traces together with unpleasure-related traces. This process permits a knowledge of external objects more or less as they are in reality, and not as if they were fragmented into two or more objects according to the specific situation they might induce in the mind at any given moment; this fragmentation is in fact the knowledge, which is erroneous, that the mind can obtain by manic operations. The two opposing categories of mental operations – the memory traces of different and opposite modes of processing the internal information of the system – are likewise put together. This can be seen as the reunification of opposing affects, the reassembly of hate and love, considered as operational schemata, and the referring of love to something internal that stems from the outside world (the object) and of hate to the subject's own internal contradictoriness. The bringing together of objects previously kept split off from each other, or of the various memory traces, so as to constitute a single memory trace that corresponds biunivocally to the external object and thus signifies it, must be complete in order for actual knowledge to arise, and for the referents to be signified to the mind by appropriate signifiers. With the completion of this process of reunification, the sign of identity (≡) in the previous formula changes to a sign of equality:

$$B_{i_g} = B_{i_b}$$

At this point, it is the child's entire self, seen as good, that is reunited with the parts formerly experienced as a bad self, thus forming a single self, which may be good or bad depending on which of its functions – affects – it is performing. This therefore calls for a process of discrimination between each of the child's own functions and, in particular, between the various functions of the child himself on the one hand and the objective attributes of real objects on the other hand. It is precisely this process that constitutes the foundation of appropriate cognitive mechanisms – appropriate not only for knowledge of the real world, but also for establishing an internal world of affects felt to belong to the subject himself, with sufficient intrapsychic permeability.

The concept of intrapsychic permeability is consistent with our definition of the unconscious in information-processing terms, as presented earlier in this book. What was formerly described as the depth of the unconscious, the greater or lesser degree of primitiveness of affects, greater or lesser conflictuality (that is, contradictory operations) and various defences (operations) – especially primitive ones such as splitting – can be redescribed in terms of the 'quality' of the internal processes, as the greater or lesser effectiveness of operations on both external and internal reality, and as a greater or lesser degree of contradiction in the operations themselves, particularly with respect to internal reality.[14] The 'deeper' unconscious is the unconscious that concerns highly contradictory and ineffective operations. Once the progressive construction of engrams (traces of more advanced functioning) allows the system to perform the operations whose subjective counterpart is some kind of capacity for consciousness, this consciousness can attain a degree of understanding of the functions that preceded it and from which it itself stems,[15] provided that the progression of signifiers is constructed harmoniously, without an excessively laborious process of overcoming contradictory operations. In other words, development is assumed to proceed by way of *chains of signifiers*, each linked in its construction to its predecessor and each determining the construction of its successor. In this construction, the progressions (or 'chains') may be ramified in more or less linear fashion, rather than being, so to speak, interwoven or mixed up, with returns to engrams of previous functions, abrupt discontinuities or voids. This may explain Bion's concept of unconscious 'lies'. I believe, too, that the 'quality' of these chains may perhaps explain the different capacities for consciousness exhibited by different subjects, or the variations in these capacities[16] observed in every individual according to the interpersonal context of the moment.

9.6 Chains of signifiers and intrapsychic permeability

Every individual has his own personal degree of being conscious of what is happening to him; in some, the so-called alexithymics, it is extremely low,

14 Some psychoanalysts may feel that the above definition impoverishes the classical psychoanalytic conception of an unconscious that is not simply not conscious, but actively kept unconscious by the force of the defences – in particular, repression. I reject this charge, which is, in my view, bound up with the stereotypical Freudian energy-and-drive conception. If the explanatory concept of energy is abandoned, as I have advocated throughout this book, there is no further reason to see the unconscious in dynamic terms. Similar considerations underlie the attempt by Schafer (1975) to eliminate all psychodynamic concepts and terminology from psychoanalytic language.

15 Cf. Bion's ideas on the role of unconscious processes in the generation of conscious thought.

16 Underlying the term 'capacity for consciousness' is a reversal of Freud's question, 'Why the unconscious?', into another – namely, 'Why consciousness?' (Imbasciati 2002b). In other words, consciousness is considered to be not something natural and constant, that can be taken for granted in any healthy adult, but the end result of a highly personal functional development depending on past and present relationships (Liotti 1994).

whereas others have a better capacity for consciousness. In analysis, the latter group are more capable of 'understanding themselves' with the help of interpretations, while the former seem deficient in this respect (Greenspan 1997), exhibiting an impermeability to interpretation (although they may oblige the analyst by appearing to accept it). In our view, the greater or lesser capacity for introspection without deception, here called the capacity for consciousness, depends on the degree of *intrapsychic permeability*, which is determined by the quality of the symbolopoietic progression of internal functional structures. The unconscious can thus be redefined in terms differing from those of the psychoanalytic tradition.

Depending on the laboriousness with which the adult subject has moved on from primitive and unrealistic operations to ones more conducive to effective knowledge, he will (especially when in analysis) have a greater or lesser capacity to achieve understanding of his unconscious mechanisms. The stage-by-stage progression from primary to more advanced operations may have taken place with a gradual or optimal connection of succeeding progressive engrams, rather than with 'returns' or regressions to primary operations (as in the case of intercurrent physical pathology or caregiver deficiencies). The chain of signifiers (Imbasciati 2001, 2002a, b, 2004b) may have been gradual and continuous, rather than interrupted and discontinuous, up to the stage of the engrams that give rise to consciousness. In our opinion, this is what determines the subject's greater or lesser capacity for consciousness (which will be greater if he is in analysis), and for being 'permeable' to the unconscious – or rather permeable *toward* his own unconscious processes. This has to do with the possibility of insight into the most remote processes, and of access to the engrams that precede the construction of those of which some kind of consciousness may be had. The more the entire progression of the various mental operations described above, at the various levels, gives rise to a gradual progression of memory traces – that is, engrams – the more this will be possible. These engrams concern the successively more advanced ways in which entities in the external world are perceived, in which they are retained in the mind rather than being expelled, and in which they are organized as memories or as the capacity to imagine something rather than to hallucinate it – as well as engrams (memory traces) of the very operational modalities that are progressively constructed on the various levels.

The construction of the mental structure, and also of what is structured by this structure (forgive the alliteration) within itself, can thus be conceived as a ramification of engrams each constructed on the basis of its predecessor. It is a ramification from the small number of internal objects, or perhaps the *only* internal object (the breast? fetal objects?), and their associated engrams to others which are both infinite and infinitely more complex, and may either be linear and gradual in its various differentiations (branches), or proceed from stage to stage with intersections, jerks, backward steps or voids. It is postulated that the factor that makes for greater or lesser intrapsychic permeability is the quality of this ramified structure: the more it is built up by gradual and linear transitions, the greater will be its continuity, which the introspective capacity (the reflexive function) constructed at the end points of some kind of consciousness will enable the individual to explore – that is to say, he will be able to have some kind of understanding (which will be truthful, rather than false and mendacious or intellectualistic) of the precursors of what appears in his mind at any given moment. This will be reflected in the individual's capacity to take advantage of analytic help.[17]

The structure of the unconscious, and indeed of each individual's specific unconscious, can therefore be described in terms of the quality of the progression of functional structures built up in the subject concerned.[18] This 'quality', or rather structure, may also be reflected in a self definable as sufficiently harmonious – rather than a false, fragmented or 'emptied-out' self – or as a degree of internal cohesion, or else as the absence or presence of lies. In the process of development of signifieds and signifiers (engrams), we have described the reunification of 'good' and 'bad' signifieds to yield a signifier denoting a single referent. This reunification, which appears self-evident for the adult mind, is by no means so in a child, in whom it entails mental suffering instead of expulsion and counter-attack. The 'suffering' may be ascribed to the 're-owning', as belonging to a 'bad' (ineffective) self, of the set of operations attributed to the object in order to avoid unpleasure. Although this 'suffering' must of course be translated into information-processing terms, this translation does not eliminate the affective semantic halo surrounding the word 'suffering': after all, the affects, which can readily be described as such in the case of more mature processes, are always the product of prior information-processing operations carried out by the system. To the adult mind, it seems natural for us to have unpleasurable things inside us, whether we are in physical pain or are afflicted by any of an infinite range of internal troubles, such as anxieties, worries, deficiencies, destructiveness, mistakes, suffering, limits or frustrations. However, this is because the adult mind has developed the strength necessary to 'cope with' these difficulties, and indeed also to withstand all the other misfortunes likely to

17 By the same token, these theoretical considerations will help the analyst to realize that resistances result from the disruption of symbolopoietic ramification, and that it is his task to reconstruct a continuity and linearity in the transitions from engram to engram, facilitating the construction of the missing engrams in the patient.

18 Our conception not only describes but also *explains* the unconscious, in terms different from the explanation afforded by Freud's energy-and-drive theory. Again, our information-processing approach – in particular, the concept of chains of symbolization – turns the traditional psychoanalytic question, 'Why the unconscious?', on its head, into the question, 'Why consciousness?', which is also more consistent with the findings of cognitivism (Imbasciati 2001, 2002a, b, 2004b).

be met with in the world. A small child's mind, by contrast, is still too rudimentary to contain its own 'deficiencies'. In other words, it cannot tolerate the mental suffering of the depressive position, which Meltzer (1979a: 132, translated) describes as in effect comprising 'a spectrum that extends from feelings of mourning, remorse and guilt, on the one hand, to those of loneliness, depression and pain due to separation, on the other'.

The achievement of this capacity is a fundamental step in development, described by psychoanalysis as the shift from the paranoid-schizoid to the depressive position. The bad object is now experienced as the subject's reaction to something unpleasurable – unpleasurable not because the object in itself is bad, but because the effect it had on the subject was painful. As and when the infant becomes in some way conscious that the person who relieves his hunger is the same person who he felt wanted to eat him, tear him to pieces and torture him, he comes to realize that his hunger, rage, soreness and other 'painful' states were not qualities of the object, but his own reaction and hence a mental function of his own. In this way, he begins to distinguish between mental objects and his own reactions to his own mental contents, and comes to discriminate between the image of a thing, the effects it may have, the circumstances in which these effects vary, and the associated emotional connotations. By successfully putting the good and bad parts together, and distinguishing the mental function from the representation of an object, the infant acquires the capacity to apply these mental functions to objects without confusion; in other words, he can then operate effectively on both internal and external reality, acting rationally in attempting to modify it. However, this entails personally owning the unpleasurable part which had previously been attributed to the outside world and had thus distorted the representation of reality: it is this recognition and recovery of the subject's own projected parts that involves pain.

Our description consistently assumes a parallelism between affects and conflictuality (the classical psychoanalytic approach), on the one hand, and, on the other hand, their logical-cognitive (involving contradiction) or information-processing counterpart, in the sense of different, succeeding and progressively more advanced ways of processing information so as to achieve the result we call knowledge of reality – whether this knowledge is of external reality or of our internal world, known with a minimum of distortion (in which case we are in *contact* with it). The latter form of knowledge cannot yet be described as consciousness, but may be called intrapsychic permeability. The parallel description of mental development in subjective-affective and in cognitive-logical-operational terms enables us to compare what have been described as painful affects, anxieties and conflicts with what we are attempting to describe in terms of the contradictoriness of mental operations. Rage, for example, is describable as a situation of contradiction between internal operations, resulting in a signal of the impossibility of effective operation. Similarly, pain and suffering can be described as rules laid down by reality, which limit the arbitrary omnipotence of mental operations.

The shift to a depressive dynamic does not mean that the earlier types of mental operations have been given up; what happens is simply that reparative operations are superimposed on them, the prevalence of the latter indicating 'normality' of mental functioning. Paranoid-schizoid and manic operations persist in adults too, but, provided that their extent is limited, are not a sign of abnormality. At unconscious level, everyone exhibits protomental as well as more advanced functioning, performing the most primitive types of mental operations. From this perspective, the shift to reparative operations signifies the beginning, in the construction of the child's mental system, of a long period of dialectical intertwining of the various dynamics, 'positions' and types of mental operations, in which one or the other type is alternately prevalent. This dialectic comes into play in so far as the mind is able to appreciate the advantages of reparative operations and to tolerate the disadvantages of the depressive position. By successfully putting together the good and the bad and hence distinguishing the subject's own mental functioning from the realistic representation of objects, the mind obtains the undoubted advantage of an increased capacity to perform effective operations on reality. However, this is accomplished at the cost of the disadvantage of having to repair the damage caused by contradictoriness – that is, of having to endure the pain of something unpleasurable, which is now experienced as belonging to the subject himself and can no longer be attributed to the outside world. If the mind does not succeed in tolerating this suffering, the individual will return to earlier paranoid-schizoid operations – but once the mind has come to know reparative operations and thus experienced the effective action of its own thoughts on reality, the former prove even more terrifying than before, because the mind now experiences their destructiveness more fully, particularly with regard to the destruction of its own thoughts, which are now felt more than ever to be the supreme asset both for the mind itself and for survival (Imbasciati 1979c). Psychotic anxieties can be understood in this light. The sense of guilt is therefore fuelled and mental suffering is exacerbated. With the advent of depressive operations, the mind not only experiences the fact of being subject to limits, to physical suffering and to dependence on others, and must not only realize that it is made up of angry, greedy and envious fantasies, as well as wishes to tear everything and everyone, including the persons and things dearest to it, to pieces, but also must ultimately take note that paranoid-schizoid operations, in which hopes of salvation and the absence of suffering had been placed, actually constitute an attack on the mind itself and tear to pieces the engrams that would otherwise have served for development, for knowledge of external reality and for obtaining relief by approaching closer to the sources of satisfaction. The paranoid-schizoid operations are thus seen to tear everything angrily to pieces, biting, lacerating and disconnecting the internal objects, whose connection, linkage and development would otherwise have served for the constitution of the mind itself and for increasing its strength.

In other words, the depressive position is in effect accompanied by a retroactive sense of guilt, which is therefore increasingly painful. The mind experiences not only the fact of being bad but also of having been wrong, of having torn itself to pieces, lacerated, gnawed at and disconnected itself; it feels devoured by rage. The sense of guilt then becomes genuine remorse – a reiterated, retroactive biting consistent with the etymology of the word 'remorse'. The two aspects – cognitive (the error) and affective (the bad fantasy) – are now united in the emerging experience that the capacity for knowledge is the true self, constitutes its nucleus, strength and development, and guarantees its survival and satisfaction.

The process of depressive reunification can be represented in mathematical-type notation as follows:

$$B_{i_g} = B_{i_b} = \text{self}$$

The sense of guilt arising from the depressive dynamic may become intolerable, in which case the mind may return to its previous form of paranoid-schizoid functioning. If the mind also has further moments of depression, this will again be experienced as terrifying, thus paradoxically increasing the terror. Hence, the oscillation between phases of reparative depressive functioning and paranoid-schizoid operation can be explained by the fact that, with the intensification of the sense of guilt in the depressive position, the paranoid-schizoid position is summoned up again unless the mind can endure the consequent increased suffering. From the reassertion of the paranoid-schizoid position, the subject must make a new attempt thus to use experiences likely to facilitate the shift to the depressive situation – only to be plunged back into the paranoid-schizoid mode of functioning once the depressive position has been attained. This is analogous to Bion's description of the Ps ↔ D oscillation. The alternation of the two positions thus takes the concrete form of reiterated attempts by the mind to 'confront' the depressive position and achieve the capacity to perform reparative (knowledge-related) operations without being thrown back by excessive guilt-based anxiety. Conversely, the store of reparative operations successfully consolidated within the mind will constitute the strength that enables it to tolerate (to know) guilt, and hence to proceed progressively along the path of these reparative operations and of development of the self – that is, of effective symbolopoiesis. In other words, there is a mental 'metabolism' that oscillates between catabolism and anabolism of structures, and which, according to the circumstances of the system's growth, may tend toward anabolism rather than annihilation.

Chapter 10

Reparation and thought

10.1 The depressive position and knowledge

The paranoid-schizoid and depressive positions can act as matrices for more advanced mental processes than those described here as protomental operations – that is to say, they can give rise to more complex functional structures, whose operationality has been described in terms of affects with which the subject relates to the world and, in particular, to the persons who can now be identified in it. These structures include what psychoanalysis has described as links or, often, 'object relationships' – that is, relationships of an affective type. The typical world of object relations in the paranoid-schizoid position appears to centre on the concern to preserve a nascent self. The object-relations world formed in the depressive position, on the other hand, lays the foundations for regarding the different-from-self as *other*, which is endowed with a particularity now recognized by the newly acquired capacity to recognize reality. The internal objects now correspond more accurately to their external counterparts, while at the same time being distinct from them – and among the external objects, *persons* (the child's caregivers) assume particular importance, while the internal objects that stand out owing to their greater intensity are those which, in affective terms, constitute links with persons. The concern to preserve the newly strengthened self can at this stage become a concern for objects, which are experienced as independent of the self. In the depressive position, this concern, although referred to the self (the 'cherished' objects), may become a concern for the object in itself, for the external object as signified by its internal counterpart, and, in particular, for the 'objects' that are persons. The foundations of the concern for the well-being, safety and happiness of the Other, and hence the goodness of the relationship with the Other, may be said to be laid at this stage. Winnicott (1958, 1965b) defines this as the phase of the 'capacity for concern' – that is, for care and solicitude. All this represents a qualitative leap in the child's mental organization. The self now presents itself as the integration of the subject's own parts and experiences and as the organizer of a dualistic world model (Money-Kyrle 1961) – that is, of a world inhabited both by a self and by other persons (objects) endowed with a psychic reality of their own, which is made up of minds and feelings.[1]

What has been described in terms of affects, or links, has likewise been recognized as the matrix for the development of thought. Depressive functioning gives rise to a fundamental development of symbolopoiesis and, in particular, to the possibility of developing the more complex functions of knowledge – that is, thought. Bion (1962, 1963, 1967, 1970) showed on the basis of clinical work that the origins of thought – including its most complex manifestations (up to the stage of conscious thought) – lie in and are modulated by the possibilities of the primitive functional levels: thought is generated by a continuous oscillation between a paranoid-schizoid and a depressive form of functioning – between Ps and D (Ps ↔ D). As Meltzer (1979a: 131, translated) notes, the two positions are

> considered to be a type of mental conflict that begins, let us say, around the third month of life and continues throughout the life of the individual. The depressive position is therefore no longer described as something that is overcome [once and for all], but as something into which the subject has 'entered' or 'penetrated'.

It is of course the entry into D, and the prevalence of D, that generate 'healthy' thought.

On this level, the child begins, albeit in rudimentary form, to experience emotions, as messages informing him of his relations with his objects; in particular, there develops a sense of guilt as a signal of the workings of his own destructiveness, or rather, of the growth of destructive-type fantasies in his mind. It is in these terms, as Fornari (1976: 131) points out, that we can understand how the self-destructive tension[2] inherent in the sense of guilt can assume the significance of an alarm signal, in the sense of representing the contradiction between destructive and self-destructive fantasies and the safety of the object of love: 'Watch out: by destroying the objects that constitute your ego (by identification), you are destroying yourself.'

The mode of organization of these internal realities is determined by the subject's fantasy life as observable in the clinical situation; having originally been expressed in the concrete form of cutting and tearing to pieces, it now involves reassembly and putting together. In this way, there emerges an

1 At this stage, the infant not only realizes that the persons around him have a psychic reality of their own, but also tends to attribute a psychic reality to inanimate objects.

2 Like other authors, Fornari considers that the sense of guilt contains a destructive impulse turned against the self (see Section 9.1, notes 2 and 3). In reality, this is so only with respect to what can appear in consciousness – that is, a self-reproach. The sense of guilt is actually self-destructive only if it is evacuated.

experience of well-being accompanied by the recognition of the existence of something good outside the confines of the subject's body. Hence, whereas the good external breast formerly consisted in not experiencing suffering, it now becomes good in itself, because it is recognized as possessing value for its own sake, by virtue of the fact of its existence. This is the result of the process of reparation achieved by the fantasies of putting things back together, which permit an actual recomposition and integration of the child's mental processes, so that he becomes capable of retaining his memory traces, repairing any damage to them and organizing them in accordance with the recognition of reality; he thus becomes truly creative. This means that thought is used efficaciously to get close to objects in reality, so that reparation coincides with the possibility of the mind's working in such a way as to cause the wished-for objects to appear in reality. Therefore, the mind ceases to be a possible channel for the entry of bad experiences[3] and becomes instead the instrument that reassures the subject of the existence of good objects in non-self reality. External reality can thus cease to be experienced as persecutory and, conversely, be felt as reassuring and possibly helpful. All this is accomplished by a process of construction of the mind, which becomes capable of mediating the possibility of knowledge; in fact, the process itself constitutes knowledge. If this process (of reparation) cannot take place, mental functioning remains on an essentially destructive and self-destructive level; hence, it can be postulated that any development of the mind must take the form of the capacity to contain the attacks on thought perpetrated by an anti-mind that effects delusional transformations (cf. Bion 1965).

The birth of thought, according to Money-Kyrle (1961; see also Meltzer 1978), involves a succession of stages, the first of which, 'subjective monism', is characterized by the use of symbolic equations (symbols are equated with their objects, or, in our terminology, signifiers are equated with signifieds and referents), in such a way that the idea of the object is equivalent to its possession. In this situation, the child's world is made up of an undifferentiated all-inclusiveness, without distinction between reality and fantasy or between inside and outside, and its spatio-temporal dimension is confined to the 'illusory present'. The use of the psychotic mechanisms of splitting and projection, and, in particular, of the mechanism of attacks on linking (Bion 1967), tends to destroy the connection between ideas and their meaning (if unpleasurable), thus hampering and delaying the differentiation between ideas and impressions and hence the use of thought to report, represent, indicate or 'symbolize' something that exists beyond the images that constitute it. At the

3 In certain psychotic structures, analysis is impeded by the fact that any contribution by the analyst, or any new experience, is felt by the analysand to be a threat, as the subject's own mind is experienced as an open door to such threats.

second stage, however, the model of the world has become 'realistic', in so far as it is felt to be made up of material objects with a permanence extending beyond the illusory present: the child is able, albeit only in embryonic form, to 'think' in terms of separate and external objects which continue to exist even if not perceptually present. However, all these objects are experienced as existing in material form[4] (expressed in our terminology, the signifier is distinguished from the signified, but the signified is fused together with the referent) and then gradually come to be seen as more 'animatistic', in the sense that the child can experience them as endowed with states of mind, which he initially projects into them, but which are later experienced as emanating from them and being directed toward himself. The intensity of projective identifications is considerable, and the mind defends against the return of threatening aspects by resorting to manic denial and, later, to repression, thus keeping at bay the possibility of perceiving the subject's own psychic reality and the image of the self as unpleasurable, greedy, angry and envious.

The third stage is described by Money-Kyrle (1961: 71) as the 'dualistic, mind-body picture of the world', and involves the capacity to use symbolic thought in relation to the internal as well as the external world:

> It is the advent of a capacity to mourn which permits the child to perceive that his present idea is not the same thing as the absent object it refers to.[5] In other words, his capacity to mourn conditions his capacity for symbolic thought. But so long as he is unable to admit his sense of responsibility for the loss of the object mourned, he dare not use his capacity for symbolic thought much on himself. For this reason, the whole inner world of his own and other people's psyche is largely closed to him. It is, then, the capacity to feel responsible for a loss mourned – and this involves the capacity to see the self from the mourned object's point of view – that conditions the development of a dualistic world-model. Of course every child who reaches the depressive position at all has in some degree attained a dualistic world-model. He thinks of himself, and other

4 In some religions, states of mind are personified in the form of actually existing entities or of deities.

5 The capacity to mourn can be understood as the realization that the subject's own representations of the world are not equivalent to the things in it, that his own internal objects cannot be equated with external objects, even if they embody them in the present or represent them: the subject's internal objects depend on the external ones, and this experience of dependence determines acceptance of the knowledge of known objects, as well as gratitude, which impresses its stamp on every true knowledge-related process, toward the objects that present themselves to be known; the subject is then prepared to depend on them even in regard to aspects of these objects that do not yield themselves up to knowledge (Imbasciati 1980).

> people, as having thoughts, feelings and desires – that is, minds – and not merely animatistic bodies.
>
> (ibid.: 72)

In other words, the signified is distinguished from the referent – the engram experienced as a signifier that bears, inside the self, the meaning of a referent. This is the actual symbolic function, which becomes an instrument for action and communication, on the one hand, and, on the other hand, for the construction of hypotheses and theories concerning, not so much reality, as the interpretation of the relationship between reality and the result of our mind's cognitive operations (Corradi Fiumara 1980). In this way, the depressive position also lays the foundations of the criterion for differentiating between misrepresentations, misconceptions and delusions, on the one hand, and scientific and cultural production, in the broad sense, on the other hand, as well as between misunderstanding and understanding, and between falsehood and truth (Money-Kyrle 1968; Bion 1970).

To sum up, the depressive position can become established in so far as the benefit – knowledge and the possibility of effective action on reality – outweighs the disadvantage of the suffering involved. It is only after the subject has embarked on this level of mental operations that both the experience of error and the sense of guilt emerge fully. The cruel fantasies that might previously have been formulated in bodily terms and hence evacuated can now be re-experienced, as if the subject had anxiously lost his bearings in recognizing himself as cruel. The self therefore emerges as such from depressive anxiety, from the feeling of having destroyed, of being destroyed, and of still being capable of destroying. The sense of guilt is now effectively a signal of death, of which the mind has learned to take account. In terms of bodily fantasies, it may take the form of having devoured the object, lacerated it, ripped it apart, torn it to pieces, consumed it and sullied it – that is, both the object loved in external reality and the object signified by the subject's own internal object. That internal object has therefore been literally 'broken' – that is to say, the cognitive structures, gradually forming or in the process of formation, have been smashed to pieces. The counterpart of this breakage is inner confusion and a return to even more destructive operations.

Depressive anxiety may thus give rise to moments of regression, involving a return to the previous level of operations, in so far as it is not endured. Alternatively, it may give rise to the need for reparation. Reparation – described by Klein by the dubious term 'impulse' – thus entails a need, or rather a necessity, to put back together or reunite what has been broken, to recreate what was destroyed. The term 'reparation' therefore readily lends itself to expressing the re-establishment of links, the putting back together of pieces, and the reinstatement of former integrity. This concerns both fantasies, which thus become capable of serving as representations of the world, and the links between them, which can therefore be reorganized in such a way as to form thought structures. What takes place on the cognitive level has its feeling equivalent in the subjective experience that emerges from this series of operations – that is, in the sense of gratification obtained by becoming capable of remaining in communication with the external objects, in the link described by the term 'gratitude' – that is love.

This concept, in Kleinian theory, denotes the feeling of recognition that links the baby to the breast and to the mother, when he recognizes that they were not bad, but that it was he himself who envied and destroyed them, and that the breast and the mother in fact offered him food and love. According to Klein and Riviere (1937), this gratitude underlies the capacity for love in both children and adults, just as the subject's admiration for his helpers enables him to learn from their experience. However, although gratitude is most obviously describable and understandable in the love relationship that may arise between two persons, it is also involved in relationships with any other external object – even, indirectly, inanimate objects. It is this gratitude that constitutes the foundation of knowledge. This has to do with the feeling that, in the construction of our internal objects – that is, of our mind – we 'depend' on external objects and on 'their' structure (Imbasciati 1980): it is not we who create objects with our thought, but objects that condition our thought. Thought is effective only if it depends on the structure of objects; we therefore accept this structure – or perhaps 'nature' – with reverence, and are grateful to our objects in so far as they have offered us this structure, instead of being angry with them because they did not allow us to create them. Offering means use, and use means knowledge. Hence, there can be no gratitude without knowledge, but likewise there can be no knowledge without gratitude: 'from learning through love to love through learning', we might say.

Reparation therefore has to do with the concrete fantasy of repairing, regenerating or creating bodies or pieces of bodies that have been destroyed in fantasy; it means putting together the destructive fantasy of breaking with the creative one, which puts the destroyed object back together with the reassembled one. All this is recognizable if these fantasies are considered in terms of the feelings which can develop in older children and in adults, and which can be attributed to infants for convenience of clinical description. In reality, the 'thoughts' of an infant differ from his more adult counterparts, which are often attributed to him. 'Destructiveness' and 'reparation' must be understood accordingly. Reparation means putting together the evacuated thought – that is, the parts of the self that had actually been smashed to pieces and destroyed – by a thought that reunifies, reassembles and seeks; that is to say, a thought that knows, a thought that actually conceives of and respects the object. This is a prerequisite for the child to learn to use a consistent thought of his own in order to obtain the things he needs, to cause them to reappear, to summon them, to ask them for something, to allow them to go away or, later, actually to construct them. In cognitive terms, this means reorganizing the memory traces – in particular, those of the various operations and functions. It may reasonably be postulated that this reorganization on the level of information processing has its counterpart in the neural network, in the form not merely of neuro-humoral changes, but also, and in particular, of an extension of the construction of the network itself, in terms of synapses and cellular proliferation.

The unification involved in the process of reparation may also remain imperfect, as it in fact usually does, in the sense that everything that can be denoted by B_{i_b} – that is, described as destructive fantasies, feelings of hate or envy, or inappropriate operations – persists within the mind, even if, in a normal mind, these remain under the control of a more appropriate organization – namely, B_{i_g}. In other words, in every human being, there persists a dialectic between thought in the narrow sense of effective and creative thought, on the one hand, and, on the other hand, destructive antimental operations, or primitive mechanisms; the outcome of this dialectic can be discerned in the capacity for truly intelligent and creative conscious thought. However, this is not the only kind of 'thought', for there can be no conscious thought without unconscious thought (Bion 1962), and no unconscious thought without the persistence of all levels of primitive mental operations. At this point, then, no thought exists without the risk of autotomic, antimental functioning – what Bion called –K. Creativity and intelligence are thus first and foremost unconscious, and can unfold in so far as the 'risk' is contained and reduced.

These dynamics can be represented by the following logic diagram:

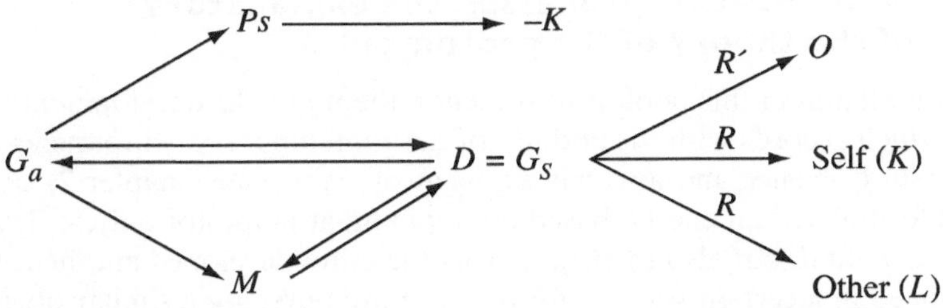

In this diagram, *G* stands for the 'guilt' that may be assumed, at primitive level, to constitute the origin of the developmental possibility of creating cognitive structures. The 'guilt', on that level, is something non-mental; it is pure aggression, indicated in the diagram by the subscript '*a*', and can be processed in the paranoid-schizoid circuits (represented by *Ps*), in which case it continues to be evacuated. Knowledge is then mere delusion (–*K*). Alternatively, it may be processed in the depressive mode (*D* in the diagram), in which case it becomes guilt-as-signal (G_s) – that is to say, it is experienced and used as such. In this case, guilt will be processed as an affect and used to safeguard and repair objects; this reparative relationship is denoted in the diagram by *R*. This promotes both the sense of a self bound up with the capacity for knowledge (*K, knowledge*) and the recognition of an external reality with which to enter into a relationship of gratitude and admiration – which, in the case of persons and, by extension, also of things, can be denoted by the recognition of an '*other*' that is loved (*L, love*). The diagram also shows another possible way, *R′*, which is less appropriate to knowledge, of processing the sense of guilt in the depressive position: this is denoted by *O*, to indicate that it has to do with obsessionality, which is characterized – precisely – by constant and defective reparation, or pseudo-reparation. Finally, the diagram also illustrates the manic situation (*M*) as a possible outcome of non-mental guilt, which, although mentalized, is, in this case, in 'hibernation' in terms of its status as a signal usable for life, unless it can be reused by a transformation into *D*, as the arrow in the diagram shows.

The diagram illustrates, too, Bion's possibility of oscillation (Ps ↔ D), but in a more specific form, which allows for the possibility that the guilt experienced on the depressive level might, if felt to be overwhelming or at any rate no longer tolerable by the mind for any length of time, once again be deprived of its mental resonance. In accordance with the notation of the diagram, it returns from G_s to G_a, and may then be re-evacuated via the paranoid-schizoid circuits (*Ps*). Alternatively, it may be encapsulated in the manic situation (*M*), from which, however, it may be revitalized in the form of depression (*D*). Finally, the diagram shows the possibility of a return from situation *D* to the manic denial of guilt (*M*).

10.2 Symbolization processes: the explanatory value of the theory of the protomental

Our overall aim in this book is to present a theory of the development of the mind which accords with the findings of the contemporary neurosciences and information science and also has explanatory value (see Chapter 2). Just as Freud formulated an energy-based theory in an attempt not only to describe and understand, but also to explain, why the mind developed and hence also functioned in a certain way, we for our part are pursuing a similar objective, albeit in a different way, with the *theory of the protomental*. As befits any explanatory theory, its status is heuristic, and not definitive. We have assumed that the neural apparatus is built to learn and that, on the basis of the form of the first inputs, it learns its own specific modes of learning, which will in turn condition the successive relationship of the nascent apparatus with subsequent experiences and its modes of learning from them. Hence the uniqueness of every individual mind and the particularity of the relations with the world that result from the possibility of learning from another thinking apparatus – that is, from interpersonal relations.

In our search for the origins of the construction of the mind, we encounter the obscure element of fetal life. This is omitted from our consideration, so that our theory, which begins only at birth, is of a provisional nature. We nevertheless believe that the theoretical framework arrived at may be useful to the student. A brief summary may be found helpful at this point.

The theory of the protomental as presented in this book explains the formation of the capacity for symbolization observed throughout the development of the mind on the basis of the aggregation of memory traces of different kinds (assemblies of afferences and efferences, as well as of traces of the progressively constructed operational modalities themselves and traces of the first 'products' of the system) into corresponding engrams that assume the status of signifiers. They signify something to the nascent system about the *world* around it, as well as about the *way* in which the system takes account of it. These distinctions are anything but simple. Although the signifiers do not at first signify anything real in either the external or the internal world, they are the first 'symbols' with which the system can begin to work. They act as units, at first quite arbitrary but gradually becoming more realistic, for reading what can be known both externally in the world and internally in its own modes of knowing. Whether the signifiers are appropriate (both for knowing the external world and for reading the internal world, as well as for not confusing the two), or are quite outlandish and absurd, they serve the purposes of the economy of the subsequent constructions built up in the mind; that is to say, they nevertheless possess symbolopoietic value.

Thus commences the symbolopoietic progression, through the chain of successive signifieds; indeed, as the chains become increasingly ramified and differentiated from engram to engram, they constitute the progressive construction of the functions of a mental structure more and more appropriate for knowledge – and hence for survival. The theory of the protomental describes anew the paranoid-schizoid position and the subsequent manic mechanisms and depressive position in terms of functions performed on the first operational units of signification and on subsequent units, which occupy an intermediate position relative to those involving representations proper. These 'intermediate representations' correspond to the developmental vicissitudes of the internal objects and the development of primary fantasies described by neo-Kleinian psychoanalysis. The various orders of primitive defences (splitting, projection-cum-evacuation, denial, idealization, and delimitation of the self from objects), up to the stage of formation of the so-called whole object, can be described in terms of operations carried out on primary representational units, from the first units that exclusively involve a dichotomy of meaning (good/bad) to other intermediate units in which signification becomes more complex. The paranoid-schizoid position is the functional matrix of the birth of perception, which is completed and stabilized in depressive functioning. The various operational stages can be schematically represented in the form of logical operations performed on the successive representational units.

Note that the various orders of successive operational units have progressively increasing representational value, defined as suitability for representing the three distinct orders of perception, memory and imagination. This representational value is indicative of a progressive development of their capacity for signification: the first elementary units simply signify the dichotomy of unpleasurable/pleasurable, which later assumes the form of the good/bad and inside/outside dichotomies. The next step, toward distinguishing between perception, memory and imagination (as well as hallucination, if the previous steps have not been successfully negotiated), entails a greater capacity for signification, with an 'inside/outside' in which reality is accommodated, and a 'self/not-self' that is also dependent on a cognition of self distinct from a cognition of the world and of others. The foundations are thus laid for the possibility of a further capacity for signification, mediated by more differentiated representations suitable for signifying ever more diverse categories, whether in perception, memory or imagination – or, in other words, by ever more complex signifiers which signify the various objects of external reality as well as, separately, the various memories and products of the subject's own mind.

Among the latter, particular attention is devoted to the products and assembled signifiers that correspond to the internal states called affects. This distinct type of signification does not yet have anything to do with consciousness.

Hence, the representational operational units constitute a ramified progression of signifiers, which the mind gradually constructs and which it can use for increasingly differentiated signifying operations. In this context, psychoanalytic cognitivism (Imbasciati 1994) sets out to describe anew the development of the symbolization processes. The entire development of the mind system, from the primary affects to the complexity of mature cognition, is illustrated in the above terms by Figure 10.1, which provides an overview of the progressive differentiation of the representational units and of the functions that elaborate and organize them. This involves a developmental process that leads, by progressive and ever more complex stages of symbolization, from affective, protomental cognition (that is, cognition mediated by the primary internal objects) to cognition proper – that is, from affect to thought.

Affects can be conceived of as a basic functional schema, or structure, of the mind system that operates as a 'reader', and regulator, of external and internal experiences, and that both underlies and conditions other functional schemata, of a cognitive-adaptive nature, that regulate behaviour, as well as the processes that may assume the character of consciousness. Affect can therefore, as Bion postulates, be said to inhere in every mental function, even when there is no obvious sign of it. There is probably a hierarchy of such basic structures, and the initial, protomental, structures may well take the form of primary internal objects, as described here. The classical psychoanalytic internal object can therefore be regarded as the primary representation of affect (Imbasciati 1991).

Its development into more advanced and complex affects, into perceptual structures rather than logical-abstract representations or into behavioural, adaptive and relational schemata, up to the stage of the appearance of, respectively, conscious perception, conscious thought and the sense of consciousness, can be conceived in terms of different forms of grouping of, and distinction between, the traces of afferences gathered and metabolized together with the signifiers progressively produced within the system.

The primary internal object takes the form of a 'set' that represents the confluence of traces of afferences of various kinds all mixed up together (for instance, buccal afferences due to stimulation by the nipple, olfactory afferences, vestibular afferences, enteroceptive afferences of hunger, other concomitant bodily afferences or possible painful stimuli), as well as of traces of other organizations built up by the initial system (ones that in the case of a more adult system would be called fantasies, experiences, feelings, imagination or thoughts). The development of this primary object into subsequent elements, or objects, more operationally appropriate to mental functioning takes place by progressive, divergent ramifications, albeit at first intersecting and interchangeable, as dictated by paranoid-schizoid functioning. Figure 10.1 takes the form of a (horizontal) ramified 'tree' commencing with the primary internal objects. The first branch comprises the separation of the two orders of traces mentioned above, resulting in the distinguishability of what will be perceivable as a sensory effect (the upper branch in the diagram) from what will be experienceable as the activity of internal functions, or of an incipient self (the lower branch).

The next ramification concerns an organization within the above two orders. In the first order (the upper branch), it leads to a separation between, and hence the distinguishability of, what is outside (and will be perceivable as an object present in external reality – in the diagram, 'external afferences' as denoted by the second downward-directed branch), and what is internal to the body (the second upward-directed branch, representing bodily afferences). With the advent of this possibility of distinction, the external afferences, of sensory origin, can be used to construct the first perceptual functions, for identifying present external objects, while the traces of afferences thus referred to the inside of the body are used no longer to signify external objects supposed to be present but, if at all, to denote absent objects and the 'need' for them – no longer, then, bad phantasms, but internal sensations. In the absence of this distinction (the up and down vertical arrows show that the separation is abolished), there is a return to the earlier confused form of functioning; this occurs in so far as paranoid-schizoid functioning becomes dominant again.

Conversely, in the second order, too (the lower branch in Figure 10.1 – representations of internal functions), a further differentiation takes place, between the various functions and activities of the self – the organization of progressively differentiated representations, of processes and of internal functions. For example, the system will be able to distinguish between a memory of something perceived and a motor memory of an interaction with the percept; between the memory of a state of bodily need and the absence of certain percepts; and, in ways that become progressively more complex, between a fantasy of the subject's own and the possibility of actively seeking out certain percepts; as well as between the construction of a spatial schema and a motor program.

The capacity to distinguish becomes progressively more subtle and differentiated toward the right of the diagram: the progressive capacity for discrimination is achieved if the primary ramification, between perceptual representations (the 'outside' – the upper branch in the diagram) and the representations of internal functions (the 'inside/self' – the lower branch), is maintained. Otherwise, if this first fundamental discrimination is lost – as indicated by the vertical arrows – paranoid-schizoid functioning once again mixes up the various orders of traces that the system had begun to distinguish. Conversely, the progressive differentiation corresponding to the depressive shift can take place by virtue of the separation – which thus permits comparison – between the various orders of afferences or traces of afferences and, in particular, between these and the various internal products, recognized as being fundamentally different from the external afferences. In this distinction, a not inconsiderable part is played by comparison of the traces from the external world with those of motor efferences: perception of the external world is distinguished by the fact that it is always accompanied by motricity, which continuously modifies it, whereas this is not the case with the perception of internal states, of bodily (biochemical) states or, still less, of products of the mind.

For this reason, Figure 10.1 illustrates the 'calibration' effected by the recording of motor efferences (E), which contributes to the distinction between outside and inside. The verification afforded by the efference, furthermore, constitutes the basis of effective interaction with reality. The calibration of E in relation to internal productions (the lower branch) is negative owing to the absence of a motor trace.

When this distinction is stably established (the vertical arrows indicating the persistence of inside/outside mixing that is characteristic of paranoid-schizoid functioning then no longer appear on the right of the diagram in Figure 10.1), the mind system can undertake an even more subtle distinction – namely, that between, on the one hand (above), the various representations that serve for perception of the outside world (mnemic categories for perceptual recognition) and, on the other hand (below), the organization of representations whereby we can read, distinguish and later possibly perceive what is happening inside ourselves (in particular, feelings, thoughts and memories). Both of these two orders of more subtle distinctions are made possible by more precise mnemic coding, which had previously been impeded by the lesser capacity of the semantic units for signification for the purposes of effective interaction with reality.

With this further functional differentiation, the system becomes capable, on the one hand, of reading the outside world, with a perception that matches reality increasingly accurately (and of no longer hallucinating by confusing outside and inside), and, on the other hand, of proceeding to refine the reading of internal reality. This involves the progressive recognition of memory and imagination, and the possibility of experiencing the many different forms whereby memories are distinguished from each other, by comparison with the even larger number of forms of imagination; in other words, the system becomes capable of reading and then possibly of perceiving[6] the internal world, the world of memories and 'images'. This last acquisition involves the possibility of the structuring of more differentiated – and more differentiable – affects (see Section 9.5). This differentiation corresponds to the construction of more and more advanced forms of symbolization through the reparative process. The progressively acquired bodily experiences will be used in this progressive process of construction (Imbasciati 1983b); progressive engrams and progressive 'modes' of functioning will be extracted from these experiences. The progressively acquired functions will then make it possible to distinguish, for example, between experiences of various kinds, between the first abstract representations and symbolizations referable to a possible concreteness of percepts – and they may also gradually allow the distinguishing of various functional connections between the subject's own mental productions and those of others, as well as of interaction with external reality, culminating in the schemata that govern interpersonal relations and the most complex adaptive behaviours, as well as the processes that entail the (progressive) understanding of what is happening not only in external reality but also within the subject himself. At this point, there are created in the system the conditions for the gradual emergence, too, of the characteristic that experiences sometimes assume which we call consciousness – the consciousness of perceiving, and hence perception of the objects of external reality, appropriate for interaction with the world; the consciousness of remembering and imagining, and hence of 'thinking' or possibly 'reasoning'; and the consciousness of feeling, or affects, which may also be conscious.

The entire developmental process involves hierarchies of differentiations in the organization of progressive representational sets. The development of affective functional structures can now be accommodated on the line of representational progression from the primary internal objects to more differentiated internal objects; various categories of the latter can be distinguished

6 'Reading' refers to a protomental organization whereby what is internal is not evacuated, but retained and used as belonging to the self, whereas 'perceiving' concerns a more advanced process that approaches closer to consciousness. An alexithymic subject can read, but not perceive, his internal world.

as differentiation progresses and an experience is formed, in accordance, precisely, with the quality of the relevant experience. One of these categories is characterized by the fact of being referred in conscious experience to objects present in external reality, and is perception; another is the consciousness of remembering something, or of thinking and reasoning, or of imagining; and yet another is referable, if it reaches the level of a conscious experience, to what we experience as, and call, a wish, shading off into what we define as a state of mind, feeling or, finally, intention. This last category includes – in a word – the affects, with their greater or lesser degree of consciousness. The order of the categories as enumerated here is, presumably, an inverse function of the degree of developmental problematicity of the internal objects involved in them, and a direct function of their representational status, up to the level of representations proper.

A continuum can thus be postulated between the internal objects proper – the primitive ones described by psychoanalysis – and representations, which serve for mature, appropriate cognition.[7] The categories set out in the final part of Figure 10.1, on the extreme right, are those 'catalogued' as such in the conscious experience of most people. There is little point in distinguishing between them before and after this level; what seems more justified is a study of the functional connections between the various orders of representations that are progressively differentiated. These functional connections underlie the interaction between affects and thought that has been, and is, central to psychoanalytic research – or rather, they underlie the interaction between the various unconscious processes themselves. In this interaction, the depressive shift indicates – as shown diagrammatically in Figure 10.1 – that the mixing of representations, or protorepresentations, belonging to the various ramifications of the differentiation is attenuated, and eventually disappears altogether when depressive functioning is fully achieved, by virtue of the consequent process of reparation and cognition.

Again, the representations located along the developmental continuum described above are not erased with the gradual progress of differentiation toward further representations, but persist as functional schemata capable of operating and of being activated to a greater or lesser degree in accordance with the circumstances of the system and the input it receives. They can be seen as the persistence of primary functions both underneath and concomitant with their more advanced secondary counterparts, in accordance with a well-known principle of psychoanalysis. The developmental continuum may be regarded as both diachronic, as in the classical tradition, and synchronic,

7 Owing precisely to the need to conceptualize the development of representations as a continuum, from affect to thought, it is unnecessary to dwell on Freud's distinctions between the various types of representations (see Section 3.6, note 7).

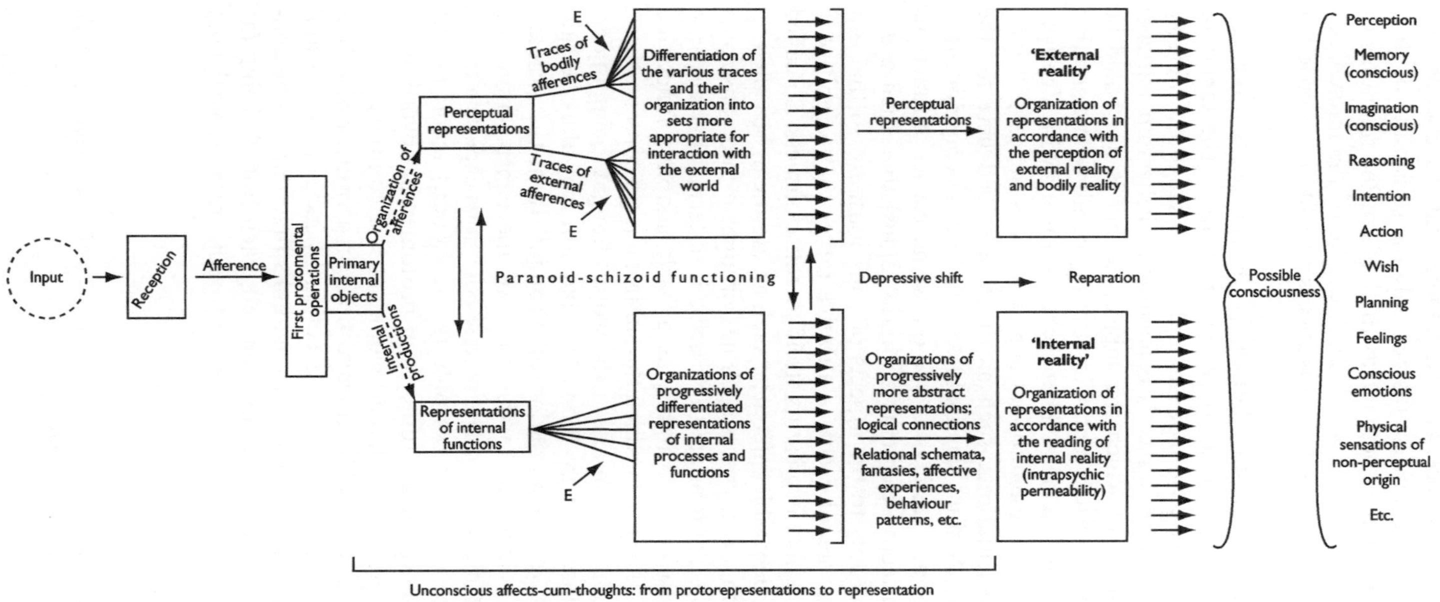

Figure 10.1 From affect to thought. This horizontal tree diagram shows the progressive ramifications (differentiations) of the sets of traces which, from the primitive internal objects, develop into more advanced experiences – that is to say, from the primary affects with their protorepresentations to representations proper. The primary internal object is made up of a particular mixture of the various traces, which, however, tend to become differentiated into two orders of percepts – namely, the experience of 'outside' (afferences) and that of memory/imagination ('inside' – that is, internal productions). The vertical arrows denote the possibility of the two orders of differentiations becoming mixed up again during the course of development, as well as in the adult unconscious. Paranoid-schizoid functioning (on the left) affords the maximum possibility of interchange between the various previously differentiated ramifications, which then become mixed together again. With the gradual yielding of paranoid-schizoid to depressive functioning, the differentiations stabilize. The calibration effected by the recording of motricity for the differentiation and organization of efferences is denoted by E (efferences): the entire course of development is governed by the interaction with reality, and motor efficacy plays a fundamental part in this interaction. A sufficient distinction between external and internal reality and an appropriate capacity for perceptual distinction between the two (intrapsychic permeability) are necessary, albeit not sufficient, conditions for awareness of the representational system on some level of consciousness.

as envisaged by Bion, in accordance with his finding that the unconscious persistence of the primary functions entails not only the possibility of their activation but also their permanent activity, as the basis of the functions of more developed 'thought'. For this reason, some degree of the interchange denoted by the vertical arrows in Figure 10.1 is always possible at unconscious level.

What appears as mature, conscious thought is the end point – the epiphenomenon – of a whole series of unconscious operations. The functions inferable from the observation of conscious mental events (like those inferable from any other observation of a subject's behaviour or behaviour pattern) involve operational schemata of which the subject neither is nor could be conscious, and which can be located on the level of the continuum extending from the primary functional schemata. These schemata work with signifiers made up of corresponding representational engrams – and hence of 'representations', or protorepresentations, such as those described here for the affects. Prior to the differentiation, classifiable on the level of consciousness, between thoughts, perceptions, feelings and so on, we have a set of functions in which cognitive and affective elements are indistinguishable: such a differentiation can be linked only to the aspects of consciousness, so that our criterion for classifying a mental event as affective rather than cognitive is the different aspect with which it presents itself to consciousness. So-called unconscious affects are a debatable inference: such an event can be called an affect only to the extent that analysis as such has brought it into consciousness.[8] Otherwise, it is more appropriate to describe it in terms that dispense with conscious subjectivity – for example, as a basic adaptive functional schema. With regard to the levels on which the subject has no consciousness, then, we may invoke a thought-cum-affect, the word 'thought' being understood in the extensive sense used by Bion. The diagram may thus be said to represent the progressive differentiation of 'thought' from the primary affects, which are unconscious, up to the point of the construction and superimposition of a possible, more or less precise conscious experience.

Note, finally, that Figure 10.1 shows only the transitions involving the greatest qualitative difference on the route from protomental operations to more complex processes. In reality, this development is a continuum of an infinite number of progressive symbolic constructions, in a symbolopoiesis in which each engram is modulated in its construction by its predecessor and in turn conditions its successor. In addition, the links that come to be constructed in the mind system are far more ramified, each branch being far more intertwined with others, than is shown in our tree diagram. In particular,

8 Cf. Freud's uncertainty on realizing that an affect, in order to be such, *must* have been conscious (Pulver 1971; Imbasciati 1991: Chapter 6).

the right-hand part of the entire diagram must be seen as much more complex, rich and ramified than shown, standing as it does for the infinitely variable capacities of human intelligence. A practically infinite number of mutually intertwined symbolic concatenations may be supposed to exist. These 'chains of signifiers' may follow straight-line paths, with linear transitions, in a progression involving a large number of intermediate steps; or, alternatively, their construction may be more contorted, with abrupt transitions, minor or even radical changes of meaning, gaps and voids in the process of signification. These may explain the lies and delusional events described by Bion. This, too, is, in my opinion, the origin of the difficulty some people have in recognizing their own internal events (alexithymia): even with the help of analysis, these subjects are unable to return to the transitions that could lead back from their consciousness to the recognition of prior processes. They have deficits (which are nevertheless cognitive) that no interpretation can make up (Greenspan 1997). Instead, a construction *ex novo* is necessary – the construction of symbolopoietic transitions, and hence new engrams, where previously there were discontinuities in symbolopoiesis. These subjects' intrapsychic permeability is blocked, and much reverie on the part of the analyst is required to reinstate it.

10.3 The maternal function and symbolopoiesis

Even if the mind system has constructed the typical functional structures of the depressive metabolism – that is, operations appropriate for the knowledge of reality and the development of thought – it continues to operate with the entire array of primitive mechanisms, although the processes initiated by the depressive position have greater intensity; at any rate, they have a greater effect on the individual's behaviour in the world, and a lesser effect in terms of internal dynamics.[9] This internal dynamic – the world of affects – remains in a state of dialectic equilibrium between the two types of mental operations; in the adult mind, a similar dialectic exists between the world of feelings and emotions and that of rationality. The oscillation described earlier (Ps $\leftrightarrow$ D) thus persists in the mind, even in normal adults. The balance and efficiency of the adult mind depend on how the oscillations between the two positions – paranoid-schizoid and depressive – became organized in early infancy. For the purposes of mental hygiene, it is therefore very important to have hypotheses about *how* these oscillations take place in the infant mind, starting from the first year of life. This is the only possible foundation for a developmental psychology that does not merely describe infant behaviour, but

9 At deeper levels, the primitive dynamics remain very active even in so-called normal persons, as is often observed in analytic treatment.

contains explanations with potential prophylactic indications. This concerns not only the alterations of development recognized as such by their pathological outcomes, but also many deficits often regarded, both in the lay mind and in scientific circles, as negligible or otherwise inevitable because they are attributed to non-mental causes. For example, greater or lesser intellective deficiencies, particularly the minor ones often considered to be within the normal range, are seen as 'natural', biological, or at any rate not due to disruption in the development of the protomental processes. Rather than developmental disturbances, these are more often regarded as 'constitutional' deficiencies. However, in the last few decades, they have been shown to be of psychogenic origin (Mannoni 1964, 1969; Bertolini, Geitlinger and Guareschi Cazzullo 1978; Bonaccorsi 1980). Particular importance has been attached to difficulties of early infantile development, with regard to the dynamics of the depressive position. As we have seen, the depressive position proves to be an essential condition for the emergence of a thinking person from an optimized process of development, as well as a fundamental determinant of the structures of human intelligence. For intelligence, as specifically manifested in *Homo sapiens*, can be traced back to the particular role of the depressive dynamic in the evolution of the human species as opposed to animals, and more especially to the longer situation of dependence (Imbasciati 1978b). This book has consistently emphasized the general implications of early infantile development for knowledge, seen in terms not of a dichotomy (intelligence versus deficit) but of a continuum. The better the development of an infant's internal world (his primary emotional structures), the more intelligent the individual in question will turn out to be. The capacity for symbolization originates here rather than in biology.

Most authors have hitherto described this development mainly in affective and emotional terms; our aim, however, has been to try to establish that this aspect is nothing other than the first, rudimentary step in cognitive development, which, as such, determines its successor. From this point of view, the process has been described in highly schematic terms: for clarity of exposition, the child has often been considered in isolation from the particularities of his interpersonal environment, and the existence of a 'standard' caregiver has been assumed. Again, the modulation of every successive step in the development of every individual by the particular elements of the interpersonal context constitutes an immense area for research. Infant psychoanalysis has to its credit a substantial number of studies that make it a discipline in its own right; however, much remains to be investigated at the earliest stages, particularly in regard to events in the minds of very small children according to their life situations. There is an even greater need to explore the fetal mind and the relationship between the pregnant mother and the fetus.

For our purposes, we need to consider how the development of the mind in the first year of life proceeds in accordance with the primary interpersonal relationship with the mother's mind,[10] which is in fact the essential modulator of cognitive development (Borgogno and Calorio 1980).

The oscillation between the paranoid-schizoid and depressive positions depends on the possibility of processing the guilt signal mentally – that is, of mentalizing it. Its mentalization entails an anxiety that can be endured only to the extent that the nascent mental apparatus is capable of enduring it, and it will be so capable only to the extent that the infant is helped by the environment (the term 'environment' is used here in Winnicott's sense, centred on the mother). It is not a matter of laying down rules and behaviours for the raising and treatment of children; such a simplistic, naive view would in any case be very confusing. What matters is the dialogue with the mother's mind, from which the baby *learns to learn from experience*; this dialogue is preverbal and based on the mother's capacity to function at protomental level. This concerns the capacity of the mother's mind for profound attunement to that of the baby, from the neonatal (and indeed fetal) stage on. Just as the developmental process in the first year of life is devoid of any element of consciousness, so the interaction between mother and child is unconscious for the mother too. She must be capable of deploying her own primitive mode of functioning in interaction with her child's mind, oscillating between the depressive and paranoid-schizoid levels. Such a process in the maternal mind is totally unconscious, and can obviously not be 'taught' in the traditional sense of the term. As Harris (1969: Section 2) points out:

> in order to understand a baby one has to feel like a baby. We have all been babies and theoretically this understanding should be within the emotional grasp of us all. But the vulnerability and acute sensibility of the baby and of the young child is something which most of us use our adult experience and competence to avoid feeling any more.

In order for the child's mental apparatus to be able to mentalize the signal of guilt, he must become capable of experiencing his own destructiveness (which, we repeat, does not mean having consciousness): this entails the child's having an experience of 'hope' that will encourage him to feel that he will become, or is becoming, capable not only of destruction but also of construction. The first element of constructiveness is afforded by the mind's own capacity to construct mental structures – effective thoughts or, if you will, good internal objects. Unless the child has such an experience sufficiently rooted in his mind, he will be unable to endure guilt, which will overwhelm and annihilate him, so that his mind will be compelled once again

10 The term 'mother' here means the regular caregiver during this period.

to evacuate it. The oscillating dialectic (Ps ↔ D) can therefore be described as a balance between despair and hope.

Despair should be seen as the inability of the mind to endure and contain guilt. Whereas this kind of language is relatively familiar in the case of adult feelings, its meaning is essentially cognitive in the mind of a child. Despair is the insufficiency of the mental apparatus for mentalizing the signal of guilt as a foundation for appropriate operations, with a consequent return to earlier operations of evacuation and autotomy. Despair can be avoided in so far as hope is present. Eschewing a description in terms of feelings (which are always more consonant with the adult than with the neonatal mind), we may define hope as the possibility that the constituted mental structures will offer a guarantee of continued survival through the process of development – that is, in the capacity to stabilize and develop good internal objects notwithstanding the presence of bad objects and, in particular, of the subject's own possible autotomic tendency. The maternal function then becomes that of sustaining the good objects and their formation.

When a small child has inside him a bad, persecutory presence of which he tries to rid himself – for instance, when he is crying so angrily and desperately that he is no longer able to do what he had seemingly already learned to do (such as waiting for the bottle, or recognizing the nipple or, later, other objects) – he attempts to expel all his badness; this he does through the voice, through convulsive movements or possibly by vomiting or excreting, or in still other ways. The mother may intervene, and her action is of fundamental importance in allowing the baby to accomplish his first 'learning' operation. In psychoanalytic parlance, the mother must be capable of receiving in her own mind the bad objects evacuated by the baby, of mentalizing them herself and of returning them to the baby in mental, and hence non-terrifying, form – that is, in the form of good, or 'thinkable', objects.

10.4 Teaching how to think: the language of reverie

The maternal function in the symbolopoietic process can be expressed as follows in terms of our theory: the mother's mind must be capable of functioning on the same protomental level as that of her baby and simultaneously on more advanced levels, with sufficient intrapsychic permeability for her to be able to move easily from one level to the other. If the child's mind is expelling bad objects, the mother should be capable of receiving the baby's signifiers (screams, movements, somatizations, etc.) and of reading their meaning. In particular, since these signifiers are fused together with the corresponding signifieds and referents, an effective maternal function should allow these 'foreign bodies' to enter and remain inside the mother, in a protomental functioning of her own, but without re-expelling them. The mother in fact attunes herself to her child's paranoid-schizoid functioning, so that the same bad objects form (enter) inside her. If she remains in her attuned position, she does the same as the child, re-evacuating the now persecutory bad object. If, on the other hand, she is capable both of this attuned functioning with the baby and of more advanced functioning of her own, she will be able to transform the bad objects, 'mentalizing' them into signifiers separate from the signifieds and referents, and at such a level that they are readable – that is, receivable – by the child's mind at that moment. She will communicate them to the child through non-verbal channels. This will enable him to accept them inside himself – no longer as solid, concrete, threatening objects, but as internal objects experienced as signifiers separable from signifieds and referents, and therefore non-persecutory (or less persecutory), because they have been detached from what has been recognized as the subject's own reaction to the object, as distinct from a property attributed to it. The child becomes capable of 'thinking' – thinking that absent objects exist, thinking his own thoughts (chagrin, disappointment, rage, aggression and envy), thinking that he can use his own reorganized mental operations to summon the object, and hence to wish and look for the object, and wait for it. In other words, the capacity of a maternal mind able to perform this function entails working in two parallel, simultaneous registers – one on the same protomental level as the child and the other on a level that is more advanced, but not excessively so, which he can reassimilate. The maternal mind works substantially with reparative processes, which act on the intolerable guilt resulting from (the child's) own destructiveness and repair the mental operations damaged by it. In order for the whole process to be effective, a good 'translation' is called for, in a 'language' the child can understand.

This maternal function of creating good internal objects, or signifiers that can be assimilated by the child's mind, is essential, for it 'teaches' the nascent mind to move on from paranoid-schizoid to depressive mechanisms, substituting for a function which the child as yet lacks and thereby teaching him to construct it. Without this maternal help, the neonatal mind will not readily be able to 'form' the first good objects, but will continue to empty itself of everything that is just beginning to form inside it. The 'mother' factor is therefore the decisive element in blocking the autotomic mechanisms and in laying the foundations for the child to move on to depressive dynamics and hence for the strengthening of his symbolic function.

This maternal function has been graphically described as the 'toilet breast', to indicate that the mother must act as a container that cleanses the child of the bad things expelled by his mind. However, since the child's mind expels everything, the mother must restore to it anything that can be recovered from his evacuative processes. The cleansing function of the maternal mind thus takes the form of accepting the bad parts, thus taking them away from the child, but also of processing them in such a way as to yield something good to give back to him – that is, offering good objects to the child's mind. Within this context of dialogue, it is therefore the mother who performs the function of gradually structuring the many and varied doings of the child, by way of her own capacity to understand his actions and types of relating and to communicate their meaning (non-verbally).

This capacity of the mother has also been called the function of 'maternal reverie', to indicate that it is performed by a mental activity resembling that of a dream (*rêve*): in other words, the mother must be capable of 'dreaming' what the child is, so to speak, dreaming. This capacity cannot be summoned up at will by a mother-to-be, but is inherent in her deep structure and, in particular, in her capacity for being more or less in contact with the most primitive parts of herself (cf. Section 9.5 on intrapsychic permeability). This capacity for 'dream-like understanding' is supported by the availability of appropriate engrams that can be conveyed in corresponding signifiers, appropriate for transmission on all the various protomental levels as and when the child acquires the capacity to receive them. In other words, the mother must be able to carry out a kind of 'simultaneous translation' from one level to another, according to the dialogue with her child. The object of the 'translation' and transmissibility is to understand the levels on which the child is 'speaking', and hence to speak to him on these or slightly higher levels, which he can nevertheless still read, thus 'teaching' him to think better and to communicate.

In cognitive terms, this means that the maternal mind 'teaches' a 'language', which structures a thought; this is of course not verbal language and verbal thought, but a type that supplies the distinction between signifiers, signifieds and referents. The neonatal mind learns to construct engrams – in particular, those referable to absent objects. In this way, it acquires for itself signifiers that signify, or have signified, referents; it confers differentiated form on what was previously formless and hence necessarily confused and ineffective for interaction with reality. It is thus a (non-verbal) language for being able to think: a thinking mind thus teaches the nascent mind to think. The mother's degree of intrapsychic permeability enables her not only to understand but also to move on from one level of language and thought to the next higher level, which, however, will be 'learnable', at that moment and in that context, by her child. In other words, the mother engages in a dialogue and teaches her baby to do the same, using increasingly advanced languages and thoughts. The child thus learns successively more mature levels of thought (see Figure 10.1).[11]

The teaching must proceed in small steps in order for learning to be possible. Having decoded a given engram of her child, the mother must offer him one that is more advanced, but not excessively so, which he can retain, or learn – that is, of which he can in turn construct the relevant engram in his own

11 This process may be likened to what Bion succinctly describes as the child's acquisition of the mother's alpha function.

mind. Mother–baby communication will be mediated by sets of afferences (which may be auditory, tactile, vestibular, visual, olfactory, etc.), suitably mixed together on each successive occasion. Their 'quality' – that is, the range of afferences – must take the form of a set readable by the child's mind at that moment; or rather, it must be *almost* readable, so that the child, in so far he can read it, can also transform it (a process involving reparation) into an engram on the next, minimally more advanced level, which will enable him to read subsequent, slightly more advanced sets of afferences. This gradual progression will depend on the mother's intrapsychic permeability, with which she can unconsciously 'teach' more or less well, so that the child will in turn 'learn' more or less easily.

The maternal function has been described classically in terms of projective and introjective identification on the part of the maternal and neonatal minds. For a specific account of these two processes, reference should be made to the psychoanalytic literature (Klein 1946; Meltzer 1973, 1979b; Rosenfeld 1980; Bion, F. 1984). In these terms, a child in the state we have described as despair may be said to be effecting a bad projective identification into the mother. She, however, will be unable to avoid introjective identification with what the baby is putting into her mind – but her mind may 'metabolize' what it receives in different ways. It may re-expel it; it may be mortally wounded (internally destroyed) by it; or it may 'understand' – in which case it will be able to give something good back to the child.

This 'understanding' means that the mother is capable of experiencing the child's anxieties, guilt and destructiveness, and of confronting them without despairing, because she has and can use efficient mental structures (good objects) that allow her to accept what is bad or 'incorrect' and to remedy it – that is, to repair it. This repair typically takes the form of a process of thought – of comprehension in the etymological sense of the word. This *comprehendere* by the mother is an acceptance and mentalization of the child's projection, which enables her to offer the child a message that he can in turn introject. The child puts referents into her, and the mother returns signifiers with signifieds to him. In this way, the child will internalize good objects and, in particular, signifiers that will allow him to repair the destruction of his mind wrought by the persecutory circuits, so that he becomes capable of 'understanding' – or, in Meltzer's term, 'digesting' – the bad experiences without being overwhelmed by them; he will thus acquire the capacity for appropriate mental operations.

In other words, if the mother 'comprehends' her own introjective identification, she will be able to effect a good projective identification in the child's mind; the neonate can 'learn' only in this way. In the same way, the mother – unconsciously – 'teaches'. This teaching, which is a teaching of signifieds, is possible in so far as the maternal capacity for reverie uses appropriate signifiers – that is to say, in so far as the mother responds with (mainly non-verbal) signifiers that can be taken in, accepted or almost perceived by the child's mind. This, then, is a very particular kind of 'teaching', mediated by an equally particular 'language'.

However, the mother's mind is not always capable of 'digesting' what the child puts into her, or of favourably metabolizing it in such a way as to return something good to him. If the mother's mind is in turn vulnerable at a deep level – that is, if her Ps ↔ D oscillation is precarious – the introduction of something bad into it will not be tolerated. In this case, the signal of guilt in the child's mind, which becomes a signal of guilt in the mother's mind, cannot be metabolized into a feeling – or rather, into mental operations – appropriate for repairing, but will be experienced by the mother with despair and terror, just as (or approximately as) the child feels it. The mother's mind will then inexorably seek to re-expel the guilt, and the child will be made the object of a bad projective identification. She will then have functioned on the same protomental level as her child, but will not have been able to complement this functioning with a more advanced functional level of her own. All the badness will therefore be returned to the baby, sometimes raised to an even higher pitch. The only course then open to him will be continued re-evacuation; his mind will thus continue to work with paranoid-schizoid mechanisms, and will be unable to learn.

These dynamics can be exemplified by two possible prototypical maternal reactions to a child's insistent crying. This may give rise to a feeling in the mother's mind such as 'Oh, poor little mite, what a bad mummy you have!', or, conversely, 'What a pest! Wait a minute, I'm coming!' As stated, these feelings are of course unconscious. The first expresses a situation of acceptance and understanding of bad phantasms, while the second indicates that the 'bad' – in this case, annoyance and irritation – is thrown back at the child by the mother. The first corresponds to the mentalization of the bad projective identification effected by the child and foreshadows the possibility of restitution of a message – a signified – conveying good objects (signifiers appropriate for knowledge), while the second indicates an expulsion by the maternal mind of the bad phantasm which the child had evacuated like a foreign body and introduced into her and which is thereby returned to the child. As a result, however good her intentions, the mother will return a bad object to the child. With the repetition of such experiences, the suffering and badness will be metabolized into something mental in the first case, whereas in the second, a permanent antimental circuit will in all probability arise between the two minds, in which case the mother will be unable to help her child's mind to develop. A mother with a type of comprehension corresponding to one of the two feelings described above will automatically transmit them to the child by way of non-verbal signals used as signifiers, so that the child's mind will, in the first case, be helped to 'understand' and hence to develop, and, in the second, be even more seriously wounded and terrified.

The above example is, of course, reductive: real dialogue situations are much more complex than the polarity described here. In all cases, however, these two opposing possibilities exist for the maternal function, with anabolic and catabolic effects respectively. Since all this occurs regardless of any conscious good intentions, but depends on the deep structure of the caregiver's mind, the importance of the mental condition of those responsible for the early care of human babies is plain, if often underestimated.

The mother's comprehension is facilitated by a mental position which accepts that the child has something bad that causes him to react to her as if she were a bad mother, accusing her of cruelty. If the mother understands how much her child is suffering, she will not feel irritated and offended by this aggression, and will thus succeed in containing it within herself and in communicating her comprehension to her child (by the tone of her voice, the touch of her hand or in other ways, which, however, will always fall outside the purview of consciousness and good intentions); this will enable him to experience the fact that he can expel his bad part into his mother without destroying anything. As Salzberger-Wittenberg (1970: 45) points out, such a mother is 'so secure in her maternal role that she is able to stand a great deal of rejection and anger from the infant and yet remain calm and lovingly concerned' – thus enabling the child to experience a differentiation between reality and fantasy.

The maternal function that promotes good introjective identifications in the child has been described (Meltzer 1973, 1979b; Bion, F. 1984) mainly in terms of the facilitation of correct emotional development, and hence as a necessary condition of the child's future mental health. The cognitive-development aspect of this function has been emphasized less frequently. The findings of Mannoni's studies (1964, 1969) of retarded children and their mothers are relevant here. This research shows that all mental retardations, from the most serious to the slightest, are of psychic origin and are due to deficiencies in mother–child communication in the first months of life. Mental deficits, which are generally attributed to established organic causes (brain disorders), can, according to this author, also be traced back to psychic variables: the organic lesions are merely predisposing and perhaps only concomitant factors.

Finally, the above considerations on general cognitive development also apply to the earliest cognitive structures – namely, those of perception. According to our hypothesis, the interplay of good identifications assumed to underlie the process described by Bion as learning from experience is already operative, and therefore observable, in the genesis of the first perceptual processes, in their early onset, their discriminatory subtlety and the possibility of the future perceptual *capacities* structuring the development of intelligence.

Similarly, according to the mother's internal mental structures and her consequent capacity to 'teach', a given type of functional structure will be transmitted to the child's mind. For instance, studies of attachment (Ainsworth, Blehar, Waters and Walls 1978; Main 1995) have demonstrated the transmission of 'internal operational models' from parents to children. It is likewise known that alexithymia results from the transmission of discontinuous chains of signifieds (Imbasciati 2004a). Hence, the mother's capacity to 'teach' is not only a quantitative but also a qualitative matter. This is the explanation of transgenerationality (Imbasciati 2002c, 2004a).

From this perspective, it is readily understandable that the mother–child relationship in the first year of the child's life is of fundamental importance for his entire future development, not only as regards the risk of pathology but also as a source of mental growth, which may be more or less rich in the individual-to-be. The mother, then, can be seen as the mediator of culture for the neonate – 'culture' here being defined as the entire mental development of the human species, both ontogenetic and phylogenetic (Imbasciati 1978b). The fundamental instrument developed by *Homo sapiens* (Imbasciati 1979c) is thus determined by the early parent–child relationship, and its most effective context is probably the family structure.

This is the fundamental and in some respects awesome background to the responsibility involved in raising children: from generation to generation, there may be either an accumulation of mental heritage, and hence of 'intelligence', or a progressive impoverishment and pathologization of minds. An important part is played here by the collective factors we call civilization – in particular, those favouring, or conversely harming, the social context in which children may be brought up.

10.5 From parent to child

Finally, when the mother's mind is the target of the infant's bad projective identification, it may respond in a third way, which can be seen as a variant of the reaction of a defective depressive response to aggression. In this case, the mother is mortally wounded and destroyed by the child's aggression; although she does not reactively throw it back to him as counter-evacuation, she is incapable of 'comprehending' it – that is, of transforming it into something mental – nor can she contain it without herself feeling her internal stability to be compromised. She may attempt to escape from the destructive feelings activated in her by the intolerable input from the baby by hurriedly offering him something to 'make him behave'. Her mind system is unable to understand what is happening in the baby's mind, so as, in turn, to make him understand it, because it cannot *comprehend*: the terror of destruction is too intense for the system and the signal of guilt arouses too much anxiety for it to be processed for the purposes of true reparation. For this reason, such a mental apparatus will effect a pseudo-reparation, obsessional in nature: such a mother will always hasten to do something for her child, but any effect this has will be only momentary. In fact, the destructiveness that caused the baby to cry will be experienced with a terror as deep as it is unconscious, so that she, like her baby, or perhaps even more so, will be incapable of tolerating the pain and fear aroused by the bad objects. Rather than repair the damage, she will be unaware of it, and will try to compensate for it by something felt to be good. But then, since there is no reparation, the search for the compensatory good will never be sufficient, and she will necessarily resort to something falsely good. She will thus deny the bad in such a way that the baby can no longer express it; for instance, she will give him anything to stop his crying, although she has not grasped its meaning and is therefore unable to teach him how to endure pain.

Such a parent, when the child is a few years old, will probably continue to give him everything, as this is her usual way of confronting aggression. However, she will in effect not be giving him anything because, by satisfying his every need to the point of satiety, she will not have enabled him to formulate an adequate wish – and the child will know nothing of himself, of what he wants, and of what is happening to him. In particular, such a parental mind will have been unable to give the child's mind what it needs most – namely, a structure that will enable the child to understand, and hence to contain and control, his own destructiveness and the effects of others' destructiveness on him. In this case, the despair is simply concealed by the fact that the child probably enjoys every kind of comfort and pleasure, every toy and every satisfaction – but will be no less profound on that account. The result may be a vicious circle in which the incapacity of the parental mind to tolerate the child's aggression will lead to his being cosseted and seduced with every kind of offer of seemingly good things, while the child's mind will not learn to tolerate anything unpleasurable, but will instead do anything to distract itself from the bad within. Being unable to suffer, the child will thus constantly provoke the parent into giving him something to make him forget that he is feeling pain, and the parent, colluding with the child's fear and obscurely perceiving, in every fresh request by the child, his terror of suffering and the fear of his own aggression, will be induced to continue this form of relationship.

What is lacking in all this is an essential quality – namely, the reparative capacity based on the acceptance of the bad, a situation involving what Meltzer (1973: 34) calls courage:

> These concepts show us adults the need to exemplify in our behaviour if we can qualities like courage, integrity and capacity for sacrifice which can be assimilated to strengthen the goodness of the child's internal objects upon whom they must, during separation and eventually, totally, depend for support.

If one's personal mental equipment lacks this capacity, any inherently good object will inexorably be consumed by destructiveness. On the basis of this dynamic between the parental unconscious and that of the child, a relationship that does not allow the formation of a sufficiently solid mental structure in the child is often established. Although he may seemingly learn to think and act intelligently, he will lack the capacity for introspection with respect to his own internal world and consequently also to the psychic reality of others; he will lack the internal creativity that permits profound and effective knowledge and, in turn, the capacity to transmit this knowledge to others by the link of love. In the absence of internal strength, constantly repeated material gifts will need to be contributed from outside (Harris and Meltzer 1980). The result may be an upbringing in which the child always has everything and can do anything he pleases. This kind of 'permissive' upbringing is often considered valid.

Now raising a child in this way may be justified and appropriate, but only as opposed to a rigid, repressive educational attitude, which may constantly and excessively impress on the child the 'bad' character of his own wishes and fantasies, thus increasing his inner terror at his own destructiveness. A rigid, repressive upbringing, then, may block creativity and hence mental development. In this sense, a more permissive upbringing may be advantageous, if the intention is to demonstrate to the developing mind that there are no mortal dangers or irremediable evils, that the mind can be exercised freely without excessive fear, and that its mistakes can be remedied. This approach, however, can easily be misunderstood, for it is substantially an unconscious inner attitude, which is on no account a matter of parental good intentions. The behavioural precept of laisser-faire, in which everything is permitted, is in fact more likely to conceal the weakness of the parent's inner structure: the resulting educational message then expresses not so much trust in the subject's own goodness and strength as a denial of his fears and despair.

In other words, the mind of a parent internally terrified by the possibility of irremediable destructiveness may deny this fear, by telling itself the lie that everything is fine and constructive and that the child can do anything freely and without excessive concerns. Such an internal economy, based on the idealization of confusion, has to do with the incapacity – described here – to tolerate destructiveness, to 'comprehend' it and to process its signal mentally, transforming it into a sense of guilt that gives rise to reparation. A parent in this situation is likely to fail to digest the bad projective identifications which the child will normally direct at him, and instead to contribute to a 'perverse' dynamic of false maturation in which the good is not sought for the purposes of reparation or creativity, but is continuously imported from the outside and inexorably burnt up. To conceal the inner reality, this permissive upbringing may be elevated to the status of an ideology – a sociological myth used as an alibi to conceal the profoundly disastrous nature of certain adult inner situations and as a defence against the failure and incapacity to establish personal relations with one's children.

In this context of a false parent–child dialogue, the permissive attitude is likely to aggravate the internal unbalance constantly. A child who is always given everything will have the obscure perception of consuming everything good without ever being capable of producing anything good himself. That is to say, the reality of his upbringing will in effect confirm to him that his destructiveness is stronger than his creativity, and that the bad in him outweighs the good. So he will become more and more terrified inside and, instead of being put on the road to the depressive position, he will be impelled toward paranoid-schizoid, or at most manic, processes. Furthermore, he will perceive that the generosity of his mother and, later, of both his parents is only apparent, the fruit not of love but of fear; he will (unconsciously) understand that they have been harmed by his destructiveness, whose uncontainability and irremediability will thereby be further confirmed. His despair will thus be raised to a even higher pitch because his parents' despair will also enter into him. His mind will consequently be unable to experience and use the sense of guilt. Instead, he will evacuate or deny every mistake and fail to recognize the true significance of his actions and feelings, so that he will inevitably develop a 'false self' (Winnicott 1965b).

Finally, if everything is permitted, the child, especially if small, will be confronted with irrefutable and utterly overwhelming proof of his own destructiveness. For instance, he will see that he can break anything he is allowed to get near, and, despite the superficial reassurance of the complaisant parent's smile (or perhaps despite his sad triumph), he is bound to be terrified at a deep level by the fact that his hands confirm what his mind already fears – namely, that he is capable only of smashing things to pieces. This view may then readily be applied to other sensory experiences too, giving many of his actions the significance of a sad confirmation of his destructive power, which he will be increasingly incapable of recognizing, or feeling guilty about, so that he will be forced more and more to increase it by denying it. In the course of his subsequent development, if a climate in which everything can be demanded and everything is ultimately granted is established with the parents, the child will experience with terror the fact of internally destroying his own parents. This event will be intolerable, and any possibility of a mental approach to it will be precluded.

In the not infrequent situation where the child, and later the adolescent, can attack the parents directly, either verbally or even physically, and they can do nothing to stem his aggression, the terror in the child's mind will be so great as to prevent him from having any consciousness of the actual meaning of his words and actions. He will thus run the risk of emptying his mind of any capacity to understand either itself or, owing to the need to rationalize his own behaviour, ideologies he may espouse to provide him with sociological, religious, philosophical or ethical explanations and justifications of his attitudes (Erikson 1968; Grinberg and Grinberg 1975; Upson 1975).

Within the limits of normal childcare, situations in which the child may experience an absence of limits to his destructiveness should therefore be avoided: besides the function of understanding the meaning of the child's actions – in this case, of the badness inside him – the parent's function is one of containment. The capacity to contain – whose prototype, for Winnicott, is the holding situation – has a mental significance that may be experienced and expressed also in physical containment of his aggression. While this aggression must not be suppressed, it can equally not be given free rein, let alone denied or minimized: it must be contained. This means that the child has an absolute need to hear the word 'no' in response to some of his demands, to see that he is prevented from smashing everything – and in particular, even if only metaphorically, everyone – to pieces. In certain circumstances, physical means – such as a salutary smack – may be the only effective way of securing understanding, in the sense that the child's mind is therefore reassured that his aggression is held in check. Such 'physical methods', however, must not have the character of a punishment – that is, of a penalty mistakenly felt to abolish guilt – but must embody the guilt itself and cause it to be used; in other words, they must constitute not retaliation but 'messages', in which the parent conveys to the child a signal that he can use.

The child needs to perceive that his 'badness' cannot prevail – that, when his mind is overwhelmed by destructiveness, someone outside himself can contain it. It is not punishment in itself that causes the child's mind to perceive that the parent is his ally against his destructive part, but the meaning that is conveyed through the physical contact. The direct and spontaneous bodily contact of such signs is often a good vehicle for a mental message signifying the containment of aggression, by which the child himself wishes to be reassured. However, it is not the action itself that is the medium of communication, but the unconscious communicative context between parent and child – and for this, it is absolutely impossible to prescribe rules of behaviour; indeed, the more considered the punishment and the more it is decided upon in cold blood, the greater the risk of conveying a contradictory message. A slap with the significance of putting a stop to the child's destructiveness is one thing, while punishments in the nature of parental revenge are another – constituting as they do instances of counter-evacuative acting out on the part of the parent, which physically visit back on the child the aggression he has projected on to the parent. Any punishment, including physical punishment, may inherently assume two opposing meanings: the first teaches the child to overcome his own badness, while the second makes him even more afraid or, in common parlance, even more 'wicked'.

The by no means negligible difference between the two types of messages that may be conveyed by the parent's strong – or rather, 'exceptional'[12] – behaviour is based on the deep echo aroused by the primary mechanisms involved in his relationship with the child. Indeed, folk wisdom – expressed in notions such as, 'I smacked you because I love you' – reflects the realization

12 Harris (1969: Section 40) writes: 'A child can come to understand the occasional hit back when he has gone too far; he can even come to understand and have some toleration for the occasions when he has not really been the guilty party but gets the blame instead of somebody else who has paved the way by too much provocation. It can even be salutary for him to realise that our patience is not inexhaustible, that we, too, as parents have rights. Salutary,

that the quality of punishment is dictated by the interpersonal meaning it conveys, which may be of love or of hate. Love will contain the destructiveness, with an understanding of the bad and of suffering, and will teach the child to tolerate it. Hate will set one destruction against another and cannot offer hope. If a child is prevented from expressing his aggression in manifestly destructive and harmful ways, this will help him to contain it as a mental state, and somehow to realize the significance of what he was doing or about to do. In so far as he expresses his rage in acts, he is unable to feel it, to conceive it in thought, and to control it. Action is the antagonist of thought; inhibiting the former encourages the latter, and vice versa. Again, as Harris (1969, Section 37) points out:

> The child cannot learn to control his undesirable, aggressive emotions unless he has had a chance to experience them, to know them at first-hand. This is the only way he can gauge their strength, the only way he can find resources within himself to harness them and, if possible, utilise them to good purpose.

However, the experience of emotions cannot take the form of outbursts of action, which in fact impede perception of the subject's emotions. The help that can be given to the child stems from the parent's capacity to set limits – on the basis not of the aspects of himself (the parent) which he cannot tolerate and has not learned to modify or integrate with the rest of his personality, and which he unconsciously experiences as projected into the child's behaviour, but of the child's needs and the harm uncontrolled destructiveness would do to the child's mental development. A mother who lets her child destroy her inflicts a terrifying experience on him, because she fails to convey any indication that would enable him to learn to control his own destructive actions, so that he is left at the mercy of unconscious fantasies of having destroyed and killed his parents. These fantasies are not only generated by the child's mind, but are due to his experience of feedback from parents who feel destroyed by him and really can allow themselves to be so destroyed. In the parents, this experience is often reflected in the reactive fantasy of killing their children, as expressed, for example, in myths and fables (Carloni and Nobili 1975; Rascovsky 1992) or acted out in criminal acts of cruelty against children, as occasionally reported in the media (Rubinstein, L.H. 1974). Apart from such extreme manifestations, it is not uncommon for parents to lack the inner strength to escape destruction by their children. Quite a few

that is, provided these spells of hitting back, physically and otherwise, occur as *exceptions* within the context of a relationship on our part that is predominantly loving, protective and secure' [my emphasis].

mothers, for example, allow themselves to be destroyed by the crying of their babies in the early months of life, and just as many no doubt react to it with hate and revenge. When a child who normally has fantasies of killing the parents sees them confirmed in whatever form by a given experience, whether of destruction or of a counter-attack that he sees he has provoked in his parents, he will be terrified by it, because he will believe that his fantasies really are omnipotent – as he once believed them to be – and really do have destructive effects on external reality. The sense of guilt will then be so annihilating that it outstrips the capacity of the depressive dynamics to cope with it, thus favouring regression to paranoid-schizoid mental states. It is thus the maternal function of containing pain, rage, guilt, uncertainty, doubt and destructiveness that promotes the development of the mind – as well as the capacity to think – because it encourages the development of

> a baby/child/adult so endowed with love that he is able to extract from experience whatever good it contains and use this to strengthen his trust, and is at the same time capable of overcoming frustrating experiences with a minimum of resentment
>
> (Salzberger-Wittenberg 1970: 44).

The secret of life thus proves to be the art of extracting good from bad – an art that parents unconsciously teach their children. It is expressed in the construction of the mind, as handed down from parents to children, and from children to their own children – for the creativity of thought is manifested in its own development of itself from generation to generation, through the translation into culture of the creativity of nature, which is the extraction of life from death. '*Bonum diffusivum sui*', said Thomas Aquinas, and the same can be said of love, defined as the development of the link, of good objects, and hence of thought, or of the Logos – or, shall we say, of the Eros-cum-Logos, affect-cum-thought, for the understanding of which our theory provides a framework.

All this, however, depends on the setting – namely, a good parent–child relationship.

References

Abercrombie, J.M. (1969) *The Anatomy of Judgement*, Harmondsworth: Penguin.
Ainsworth, M., Blehar, M., Waters, E. and Walls, S. (1978) *Patterns of Attachment. A Psychological Study of the Strange Situation*, Hillsdale, NJ: Erlbaum.
Alexander, F. (1956) *Psychoanalysis and Psychotherapy*, New York: Norton.
Allport, F.H. (1955) *Theories of Perception and the Concept of Structure*, New York: Wiley.
Argyle, M. (1975) *Bodily Communication*, London: Methuen.
Arieti, S. (1969) *The Intrapsychic Self*, New York: Basic Books.
Arnheim, R. (1970) *Visual Thinking*, London: Faber and Faber.
—— (1974) *Art and Visual Perception*, Berkeley, CA: University of California Press.
Arnold, W., Eysenck, H.J. and Meili, R. (1972) *Encyclopedia of Psychology*, London: Search Press.
Benedetti, G. (1969) *Neuropsicologia*, Milan: Feltrinelli.
Bernfeld, S. (1929) [1925] *The Psychology of the Infant*, trans. R. Hurwitz, London: Kegan, Paul.
Bertolini, M., Geitlinger, G. and Guareschi Cazzullo, A. (1978) 'Normalità, salute e malattia nel bambino', *Tempo Medico*, 4: 3–10.
Bick, E. (1964) 'Notes on infant observation in psycho-analytic training', *International Journal of Psychoanalysis*, 45: 558–66.
—— (1968) 'The experience of the skin in early object relations', *International Journal of Psychoanalysis*, 49: 484–6.
Bieri, J., Atkins, A.L., Briar, S. *et al.* (1966) *Clinical and Social Judgement: The Discrimination of Behavioral Information*, New York: Wiley.
Bion, F. (ed.) (1984) *Discussioni con Bion*, Turin: Loescher [Italian compilation containing Bion's *Four Discussions with W.R. Bion* and *Bion in New York and São Paulo*].
Bion, W.R. (1962) *Learning from Experience*, London: Heinemann.
—— (1963) *Elements of Psycho-Analysis*, London: Heinemann.
—— (1965) *Transformations*, London: Heinemann.
—— (1967) *Second Thoughts*, London: Heinemann.
—— (1970) *Attention and Interpretation*, London: Tavistock.
—— (1974) *Il cambiamento catastrofico*, Turin: Loescher, 1981 [Italian compilation containing 'Catastrophic change' and the 1976 Brazilian Lectures].
Bleger, J. (1967) *Simbiosis y ambigüedad*, Buenos Aires: Paidós.

Boddy, J. (1978) *Brain Systems and Psychological Concepts*, Chichester: Wiley.
Bollas, C. (1987) *The Shadow of the Object*, London: Free Association Books.
—— (1992) *Being a Character*, New York: Giroux.
Bonaccorsi, M. (1980) *La psicoterapia psicoanalitica del bambino organico*, Milan: Emme.
Borgogno, F. (1978) *L'illusione di osservare*, Turin: Giappichelli.
—— and Calorio, D. (1980) 'Considerazioni sullo studio delle basi dello sviluppo cognitivo', in P. Amerio and G.P. Quaglino, *Mente e società nella ricerca psicologica*, Turin: Book Store.
Bower, T.G.R. (1974) *Development in Infancy*, San Francisco, CA: Freeman.
Bowlby, J. (1969, 1973, 1980) *Attachment and Loss* (vols 1, 2 and 3), New York: Basic Books.
—— (1979) *The Making and Breaking of Affectional Bonds*, London: Tavistock.
Brown, J. (1977) *Mind, Brain and Consciousness. The Neuropsychology of Cognition*, New York: Academic Press.
Brown, R. (1973) *A First Language*, Cambridge, MA: Harvard University Press.
Bruner, J.S. (1968) *Processes of Cognitive Growth: Infancy*, Worcester, MA: Clark University Press.
Brunswik, E. (1956) *Perception and the Representative Design of Psychological Experiments*, Berkeley, CA: University of California Press.
Bucci, W. (1997) *Psychoanalysis and Cognitive Sciences*, New York: Guilford.
Calorio, D. and Purghé, F. (1977) 'Le illusioni ottico-geometriche: contributo sperimentale alla teoria della gradazione della costanza di Gregory', *Bollettino di Psicologia Applicata*, 139: 145–80.
Camaioni, L. (1993) *La teoria della mente*, Bari: Laterza.
Camaioni, L., Volterra, V. and Bates, E. (1976) *La comunicazione nel primo anno di vita*, Turin: Boringhieri.
Carloni, G. and Nobili, D. (1975) *La mamma cattiva. Fenomenologia e antropologia del figlicidio*, Florence: Guaraldi.
Carterette, E.C. and Friedman, M.P. (1974–8) *Handbook of Perception*, New York: Academic Press, vol. 1, *Historical and Philosophical Roots of Perception* (1974); vol. 4, *Hearing* (1978a); vol. 5, *Seeing* (1975); vol. 6a, *Tasting and Smelling* (1978b); vol. 6b, *Feeling and Hurting* (1978c); vol. 7, *Language and Speech* (1976); vol. 8, *Perceptual Coding* (1978d); vol. 9, *Perceptual Processing* (1978e).
Corradi Fiumara, G. (1980) *Funzione simbolica e filosofia del linguaggio*, Turin: Boringhieri.
Cromer, R.F. (1974) 'The development of language and cognition', in B.M. Foss, *New Perspectives in Child Development*, Harmondsworth: Penguin, 184–252.
Davidson, J.M. and Davidson, R.J. (1980) *The Psychobiology of Consciousness*, New York: Plenum Press.
Della Vedova, A. (1999) 'La vita psichica prenatale'. *Psychomedia*, 23 February 1999. www.psychomedia.it/pm/lifecycle/perinatal/delved1.htm (accessed 14 May 2004).
—— and Imbasciati, A. (1998) 'Alle origini della mente', *Giornale di Neuropsichiatria dell'età evolutiva*, 18: 155–68.
——, Manfredi, P. and Imbasciati, A. (1996) 'La comunicazione auditiva gestante-feto'. *Psychomedia*, November. Online. www.psychomedia.it/pm/lifecycle/perinatal/imbasc1a.htm (accessed 14 May 2004).

Dufoyer, J.-P. (1976) *Le Développement psychologique de l'enfant de 0 à 1 an*, Paris: Presses Universitaires de France.

Eagle, M. (1984) *Recent Developments in Psychoanalysis: A Critical Evaluation*, New York: McGraw-Hill.

Eccles, J.C. (1970) *Facing Reality. Philosophical Adventures by a Brain Scientist*, London: Longman.

Ehrenzweig, A. (1967) *The Hidden Order of Art*, London: Paladin.

—— (1975) *The Psychoanalysis of Artistic Vision and Hearing*, London: Sheldon Press.

Eiser, J.R. and Stroebe, W. (1972) *Categorization and Social Judgement*, London: Academic Press.

Erikson, E.H. (1950) *Childhood and Society*, New York: Norton.

—— (1968) *Identity, Youth and Crisis*, New York: Norton.

Fabozzi, P. and Ortu, F. (1996) *Al di là della metapsicologia*, Rome: Il Pensiero Scientifico.

Fairbairn, W.R.D. (1952) *Psychoanalytic Studies of the Personality*, London: Tavistock.

Fantz, R.L. (1965) 'Ontogeny of perception', in A.M. Schrier, H.F. Harlow and F. Stollnitz, *Behavior of Nonhuman Primates*, New York: Academic Press.

Fornari, F. (1971) *La vita affettiva originaria del bambino*, Milan: Feltrinelli.

—— (1975) *Genitalità e cultura*, Milan: Feltrinelli.

—— (1976) *Simbolo e codice*, Milan: Feltrinelli.

Foss, B.M. (1961) *Determinants of Infant Behaviour*, London: Methuen.

—— (ed.) (1974) *New Perspectives in Child Development*, Harmondsworth: Penguin.

Fossi, G. (1984) *Le teorie psicoanalitiche*, Padua: Piccin.

Freedman, N. and Grand, S. (1977) *Communicative Structures and Psychic Structures*, New York: Plenum Press.

Freud, A. (1965) *Normality and Pathology in Childhood*, London: Hogarth.

Freud, S. (1895d [1893–95]) (with J. Breuer) *Studies on Hysteria, Standard Edition*, 2.

—— (1905d) *Three Essays on the Theory of Sexuality, Standard Edition*, 7.

—— (1905e [1901]) 'Fragment of an analysis of a case of hysteria', *Standard Edition*, 7.

—— (1906a) 'My views on the part played by sexuality in the aetiology of the neuroses', *Standard Edition*, 7.

—— (1914c) 'On narcissism: an introduction', *Standard Edition*, 14.

—— (1915c) 'Instincts and their vicissitudes', *Standard Edition*, 14.

—— (1916–17a [1915–17]) *Introductory Lectures on Psycho-Analysis, Standard Edition*, 15–16.

—— (1920g) *Beyond the Pleasure Principle, Standard Edition*, 18.

—— (1923b) *The Ego and the Id, Standard Edition*, 19.

—— (1923e) 'The infantile genital organization', *Standard Edition*, 19.

—— (1924c) 'The economic problem of masochism', *Standard Edition*, 19.

—— (1924d) 'The dissolution of the Oedipus complex', *Standard Edition*, 19.

—— (1925j) 'Some psychical consequences of the anatomical distinction between the sexes', *Standard Edition*, 19.

—— (1933a [1932]) *New Introductory Lectures on Psycho-Analysis, Standard Edition*, 22.

—— (1937c) 'Analysis terminable and interminable', *Standard Edition*, 23.

—— (1985c [1887–1904]) *The Complete Letters of Sigmund Freud to Wilhelm Fliess*

1887–1904, trans. and ed. J.M. Masson, Cambridge, MA: Belknap Press of Harvard University Press.

Frisch, K. von (1938) [1933] *The Language of Bees*, Annual Report of the Board of Regents of the Smithsonian Institution, Washington, DC: US Government Printing Office.

Gaddini, R. (1978) 'Transitional object origins and the psychosomatic symptom', in S.A. Grolnick and L. Barkin (eds), *Between Reality and Fantasy*, New York: Aronson.

Gedo, J.E. and Goldberg, A. (1973) *Models of the Mind. A Psychoanalytic Theory*. Chicago: University of Chicago Press.

Ghilardi, A. and Imbasciati, A. (1990) 'L'istinto nelle scienze: necessità e storia di un concetto irresolubile', *Rivista di Sessuologia*, 14: 101–16.

Gill, M.M. (1976) 'Metapsychology is not psychology', in M.M. Gill and P.S. Holzman (eds), 'Psychology versus metapsychology. Psychoanalytic essays in memory of George S. Klein', *Psychological Issues*, 9, Monograph 36, New York: International Universities Press.

Gilli, A. and Marchetti, L. (1992) *Processi sociocognitivi*, Milan: Raffaello Cortina.

Girotti, G. and Calorio, D. (1974) 'Determinazioni sperimentali sull'illusione di Muller-Lyer nell'adolescenza', *Contributi dell'Istituto di Psicologia*, 32, Milan: Vita e Pensiero.

Gori, C.E. (1991) *Psicoanalisi e processo creativo*, Milan: Angeli.

Greenberg, J. and Mitchell, S. (1983) *Object Relations in Psychoanalytic Theory*, Cambridge, MA: Harvard University Press.

Greenspan, S. (1997) *Developmentally Based Psychotherapy*, New York: McGraw-Hill.

Grinberg, L. (1992) [1976] *Guilt and Depression*, trans. C. Trollope, London: Karnac. 1992.

Grinberg, L. and Grinberg, R. (1975) *Identitad y cambio*, Buenos Aires: Editorial Nova.

Guntrip, H. (1961) *Personality Structure and Human Interaction*, New York: International Universities Press.

Hamilton, V. and Vernon, M. (1976) *The Development of Cognitive Processes*, London: Academic Press.

Harlow, H.F. (1961) 'The development of affectional patterns in infant monkeys', in B.M. Foss (ed.), *Determinants of Infant Behaviour*, vol. 1, London: Methuen.

Harlow, H.F. and Harlow, M.K. (1965) 'The affectional systems', in A.M. Schrier, H.F. Harlow and F. Stollnitz, *Behaviour of Nonhuman Primates*, London: Academic Press.

Harris, M. (1969) *Inside Information on Understanding Infants (Birth to Five Years)*, London: Dickens Press.

—— and Meltzer, D. (1980) *A Psychoanalytical Model of the Child-in-the-Family-in-the-Community*, in D. Meltzer, *Sincerity and Other Works*, London: Karnac, 1994.

Hartmann, H. (1939) *Ego Psychology and the Problem of Adaptation*, trans. D. Rapaport, London: Imago, 1958.

Hebb, D.O. (1949) *The Organization of Behavior*, New York: Wiley.

Heimann, P. (1952) 'Notes on the theory of the life and death instincts', in J. Riviere (ed.), *Developments in Psycho-Analysis*, London: Hogarth.

Hinde, R.H. (1974) *Biological Bases of Human Social Behaviour*, New York: McGraw-Hill.

Holt, R.R. (1972) 'Freud's mechanistic and humanistic image of man', *Psychoanalysis and Contemporary Science*, 1: 3–24.
Hoxter, S. (1977) 'Play and communication', in M. Boston and D. Daws (eds), *The Child Psychotherapist*, London: Wildwood House.
Hutt, S.J. and Hutt, C. (1973) *Early Human Development*, London: Oxford University Press.
Hutten, E.H. (1962) *The Origins of Science*, London: Allen and Unwin.
Imbasciati, A. (1967) 'Il concetto di proiezione', *Archivio di Psicologia, Neurologia e Psichiatria*, 28: 170–82.
—— (1978a) *Principi introduttivi alla psicoanalisi*, Milan: Angeli.
—— (1978b) *Eros e Logos: amore, sessualità e cultura nello sviluppo dello spirito umano*, Brescia: La Scuola.
—— (1979a) 'Gyné, la creatura che crea', in *Psicoanalisi e femminilità*, Milan: Angeli.
—— (1979b) 'Il senso di colpa nella sessualità', *Rivista di Psicoanalisi*, 25: 257–81.
—— (1979c) 'L'ansia epistemofilica', *Gli Argonauti*, 1: 175–87.
—— (1980) Foreword (in Italian) to G. Morino, *Il concetto di identificazione*, Turin: Giappichelli.
—— (1983a) *Sviluppo psicosessuale e sviluppo cognitivo*, Rome: Il Pensiero Scientifico.
—— (1983b) 'L'intelligenza del sogno', *Il Ragguaglio Librario*, 50: 281–3.
—— (1983c) *Freud o Klein? La femminilità nella letteratura psicoanalitica*, Rome: Armando.
—— (1986) *Istituzioni di psicologia*, 2 vols, Turin: Utet Libreria.
—— (1987) 'Un modello psicoanalitico dei processi cognitivi', *Professione Psicologo*, 4: 21–33.
—— (1989a) 'Towards a psychoanalytic model of cognitive processes', *International Review of Psychoanalysis*, 16: 223–36.
—— (1989b) *La consapevolezza*, Rome: Borla.
—— (1990) *La donna e la bambina*, Milan: Angeli.
—— (1991) *Affetto e rappresentazione*, Milan: Angeli.
—— (1993a) *Psicologia medica*, Naples: Idelson Liviana.
—— (1993b) *L'oggetto e le sue vicissitudini*, Castrovillari: Teda.
—— (1994) *Fondamenti psicoanalitici della psicologia clinica*, Turin: Utet Libreria.
—— (1996) 'Comunicazione gestante-feto e madre-bambino e sviluppo psichico del bambino', *PsychoMedia*, November. Online. www.psychomedia.it/pm/lifecycle/perinatal/imbasc1a.htm (accessed 9 June 2004).
—— (1997) 'Comunicazione gestante e feto e madre-neonato e sviluppo psichico e psicosomatico del bambino', in C. Bosio (ed.), *Psicologia medica*, Milan: Angeli.
—— (1998) *Nascita e costruzione della mente*, Turin: Utet Libreria.
—— (2000) 'Le origini della dimensione sessuale', *Rivista di Sessuologia*, 24: 149–61.
—— (2001) 'The unconscious as symbolopoiesis', *Psychoanalytic Review*, 88: 837–73.
—— (2002a) 'An explanatory theory for psychoanalysis', *International Forum of Psychoanalysis*, 11: 173–83.
—— (2002b) 'A psychoanalyst's reflections on rereading a cognitivist', *Psychoanalytic Review*, 89: 595–630.
—— (2002c) 'Un supporto teorico alla transgenerazionalità', *Atti IV Congresso Psicologia Prenatale*, Rome, Brescia: Cantagalli, 2003, 31–48.
—— (2003) 'Cognitive sciences and psychoanalysis: a possible convergence', *Journal of the American Academy of Psychoanalysis*, 31: 627–43.

Imbasciati, A. (2004a) 'A theoretical support for transgenerationality: the theory of the protomental', *Psychoanalytic Psychology*, 21: 83–98.

—— (2004b) *Compendio di psicologia per gli operatori socio-sanitari*, Padua: Piccin.

—— and Calorio, D. (1981) *Il Protomentale*, Turin: Boringhieri.

—— and Ghilardi, A. (1989) 'Il concetto di istinto e il suo uso in psicoanalisi', *N.P.S. Neurologia, Psichiatria, Scienze Umane*, 9: 1035–56.

Isaacs, S. (1948) 'The nature and function of phantasy', in M. Klein, P. Heimann, S. Isaacs and J. Riviere, *Developments in Psycho-Analysis*, London: Hogarth, 1952.

Josselyn, I.M. (1971) *Adolescence; a Report Published Under the Auspices of the Joint Commission on Mental Health of Children*, New York: Harper and Row.

Jouvet, M. (1978) 'Does a genetic programming of the brain occur during paradoxical sleep?', in P.A. Buser and A. Rougeul-Buser, *International Symposium on Cerebral Correlates of Conscious Experience*, Amsterdam: North-Holland.

Kanizsa, G. (1968) 'Percezione attuale, esperienza passata e "l'esperimento impossibile" ', in G. Kanizsa and G. Vicario, *Ricerche sperimentali sulla percezione*, Università degli Studi di Trieste.

—— (1975) 'La percezione', in G. Kanizsa, P. Legrenzi and P. Meazzini, *I processi cognitivi*, Bologna: Il Mulino.

—— (1980) *Grammatica del vedere*, Bologna, Il Mulino.

—— and Legrenzi, P. (1978) *Psicologia della Gestalt e psicologia cognitivistica*, Bologna: Il Mulino.

Katz, D. (1951) *Gestalt Psychology: Its Nature and Significance*, London: Methuen.

Kaufman, L. (1974) *Sight and Mind*, New York: Oxford University Press.

Kellerman, H. (1980) 'An epigenetic theory of emotion in early development', in R. Plutchik and H. Kellerman (eds), *Emotion, Theory, Research and Experience*, New York: Academic Press, 1980–5, 3 vols.

Kelly, G.A. (1955) *The Psychology of Personal Constructs*, New York: Norton.

Klein, G. (1976) *Psychoanalytic Theory: An Exploration of Essentials*, New York: International Universities Press.

Klein, M. (1930) 'The importance of symbol formation in the development of the ego', in *Contributions to Psycho-Analysis 1921–1945*, London: Hogarth.

—— (1932) *The Psycho-Analysis of Children*, London: Hogarth.

—— (1935) 'A contribution to the psychogenesis of manic-depressive states', in *Contributions to Psycho-Analysis 1921–1945*, London: Hogarth.

—— (1940) 'Mourning and its relation to manic-depressive states', in *Contributions to Psycho-Analysis 1921–1945*, London: Hogarth.

—— (1945) 'The Oedipus complex in the light of early anxieties', in *Contributions to Psycho-Analysis 1921–1945*, London: Hogarth.

—— (1946) 'Notes on some schizoid mechanisms', in *The Writings of Melanie Klein*, vol. 3, London: Hogarth, 1975.

—— (1948) 'On the theory of anxiety and guilt', in *The Writings of Melanie Klein*, vol. 3, London: Hogarth, 1975.

—— (1952) 'Some theoretical conclusions regarding the emotional life of the infant', in *The Writings of Melanie Klein*, vol. 3, London: Hogarth, 1975.

—— (1957) 'Envy and gratitude. A study of unconscious sources', in *The Writings of Melanie Klein*, vol. 3, London: Hogarth, 1975.

—— (1960) 'A note on depression in the schizophrenic', *International Journal of Psychoanalysis*, 41: 247–55.

—— (1961) 'Narrative of a child analysis', in *The Writings of Melanie Klein*, vol. 4, London: Hogarth, 1975.
—— (1963) *Our Adult World and Other Essays*, London: Heinemann.
—— (1973) *Contributions to Psycho-Analysis 1921–1945*, London: Hogarth.
—— (1975) *The Writings of Melanie Klein*, 4 vols, London: Hogarth.
—— and Riviere, J. (1937) *Love, Hate and Reparation*, London: Hogarth.
Koffka, D. (1935) *Principles of Gestalt Psychology*, New York: Harcourt Brace.
Köhler, W. (1929) *Gestalt Psychology*, New York: Liveright.
Kohut, H. (1971) *The Analysis of the Self*, New York: International Universities Press.
—— (1977) *The Restoration of the Self*, New York: International Universities Press.
Kreisler, L., Fain, M. and Soulé, M. (1974) *L'Enfant et son corps*, Paris: Presses Universitaires de France.
Laplanche, J. and Pontalis, J.-B. (1964) 'Fantasme originaire, fantasmes des origines, origine du fantasme', *Les Temps Modernes*, 215: 1833–68.
—— (1967). *The Language of Psycho-Analysis*, trans. D. Nicholson-Smith, London: Hogarth, 1973.
Lebovici, S. and Kestemberg, E. (1978) *Le Devenir de la psychose de l'enfant*, Paris: Presses Universitaires de France.
Lichtenberg, J.D. (1989) *Psychoanalysis and Motivation*, Hillsdale, NJ: Analytic Press.
Liotti, G. (1994) *La dimensione interpersonale della coscienza*, Rome: NIS.
Lorenz, K. (1965) *Evolution and Modification of Behaviour*, Chicago: University of Chicago Press.
—— (1969) 'Innate bases of learning', in K.H. Pribram (ed.), *On the Biology of Learning*, New York: Harcourt Brace.
Lunzer, E.A. (1979) 'The development of consciousness', in G. Underwood and R. Stevens, *Aspects of Consciousness*, vol. 1, London: Academic Press.
Luria, A.R. (1974) [1970] *Neuropsicologia e neurolinguistica*, Rome: Editori Riuniti (Italian compilation).
—— (1974) *The Working Brain: An Introduction to Neuropsychology*, trans. B. Haigh, New York: Basic Books.
—— (1976) *Basic Problems of Neurolinguistics*, The Hague: Mouton.
Mahler, M.S., Pine, F. and Bergman, A. (1975) *The Psychological Birth of the Human Infant*, New York: Basic Books.
Main, M. (1995) 'Recent studies in attachment', in S. Goldberg, R. Muir and J. Kerr (eds), *Attachment Theory: Social, Developmental and Clinical Perspectives*, Hillsdale, NJ: Analytic Press.
Mancia, M. (1980) *Neurofisiologia e vita mentale*, Bologna: Zanichelli.
Manfredi, P. and Imbasciati, A. (1997) 'La percezione acustica fetale', *Archivio di Psicologia, Neurologia e Psichiatria*, 58: 165–88.
Manfredi, P., Imbasciati, A. and Ghilardi, A. (1997) 'L'esposizione auditiva fetale: uno strumento di indagine', *Imago*, 4: 89–106.
Manfredi, P., Tommasoni, V. and Imbasciati, A. (1999) 'L'esposizione auditiva fetale: una ricerca sulle madri in gravidanza', *Psicologia Clinica dello Sviluppo*, 3: 37–60.
Mannoni, M. (1964) *L'Enfant arriéré et sa mère*, Paris: Seuil.
—— (1969) *L'Enfant, sa maladie et les autres*, Paris: Payot.
Marcuse, H. (1955) *Eros and Civilization: A Philosophical Inquiry into Freud*, Boston: Bacon Press.
—— (1960) *Studie über Autorität und Familie*, Frankfurt am Main: Suhrkamp.

Maturana, H. and Varela, F. (1980) *Autopoiesis and cognition*, Dordrecht: Reidel.
McGurk, H. (1974) 'Visual perception in young infants', in B.M. Foss (ed.), *New Perspectives in Child Development*, Harmondsworth: Penguin.
Mehrabian, A. (1971) *Silent Messages*, Belmont: Wadsworth.
Meltzer, D. (1973) *Sexual States of Mind*, Perthshire: Clunie Press.
—— (1978) *The Kleinian Development*: Part I, *Freud's Clinical Development*; Part II, *Richard-Week-by-Week*; Part III, *The Clinical Significance of the Work of Bion*, Perthshire: Clunie Press.
—— (1979a) 'Posizione schizoparanoide e posizione depressiva', *Quaderni di Psicoterapia Infantile*, No. 1: 125–41.
—— (1979b) 'Sul processo introiettivo', *Quaderni di Psicoterapia Infantile*, No. 2: 219–59.
——, Bremner, J., Hoxter, S., Weddell, D. and Wittenberg, I. (1975) *Explorations in Autism: A Psycho-Analytical Study*, Perthshire: Clunie Press.
—— and Harris, M. (1986) *Il ruolo educativo della famiglia*, Turin: Centro Scientifico Torinese (Italian compilation).
Metzger, W. (1954) *Psychologie: Die Entwicklung ihrer Grundannahmen seit der Einführung des Experiments*, Darmstadt: Steinkopff.
Miller, G.A. (1967) *The Psychology of Communication*, New York: Basic Books.
Modell, A. (1975a) 'The ego and the id: 50 years later', *International Journal of Psychoanalysis*, 56: 57–68.
—— (1975b) *Object Love and Reality*, New York: International Universities Press.
Money-Kyrle, R. (1955) 'An inconclusive contribution to the theory of death instinct', in M. Klein, P. Heimann and R. Money-Kyrle (eds), *New Directions in Psycho-Analysis*, London: Tavistock, 409–59.
—— (1961) *Man's Picture of His World: A Psycho-Analytic Study*, London: Duckworth.
—— (1978) [1956] 'Psycho-analysis and philosophy', in R. Money-Kyrle 1978, 297–317.
—— (1978) [1965] 'Success and failure in mental maturations', in R. Money-Kyrle 1978, 397–406.
—— (1978) [1968] 'Cognitive development', in R. Money-Kyrle 1978, 416–33.
—— (1978) *The Collected Papers of Roger Money-Kyrle*, D. Meltzer and E. O'Shaughnessy (eds), Perthshire: Clunie Press.
Morino, G. (1980) *Il concetto di identificazione*, Turin: Giappichelli.
Nathanielsz, P.W. (1994) *A Time to Be Born: The Life of the Unborn Child*, Oxford: Oxford University Press.
Negri, R. (1993) *Il neonato in terapia intensiva*, Milan: Raffaello Cortina.
Neisser, U. (1967) *Cognitive Psychology*, New York: Appleton-Century-Crofts.
—— (1976) *Cognition and Reality*, San Francisco, CA: Freeman.
Ogden, T.H. (1990) *The Matrix of the Mind*, Northvale, NJ: Jason Aronson.
Oliverio, A. (1986) *Storia naturale della mente*, Turin: Bollati Boringhieri.
Peterfreund, E. (1971) *Information, Systems and Psychoanalysis*, New York: International Universities Press.
—— and Franceschini, E. (1973) 'On information, motivation and meaning', *Psychoanalysis and Contemporary Science*, 2: 220–62.
Piaget, J. (1969) *The Mechanisms of Perception*, trans. G.N. Seagrim, London: Routledge and Kegan Paul.

Piontelli, A. (1987) 'Infant observation from before birth', *International Journal of Psychoanalysis*, 68: 453–63.

Plutchik, R. (1980) 'A general psychoevolutionary theory of emotion', in R. Plutchik and H. Kellerman (eds), *Emotion, Theory, Research and Experience*, vol. 1, New York: Academic Press.

Pulver, S. (1971) 'Can affects be unconscious?', *International Journal of Psychoanalysis*, 52: 347–54.

Purghé, F. and Imbasciati, A. (1979) *Psicologia dei processi visivi*, Rome: Il Pensiero Scientifico.

Rascovsky, A. (1995) [1992] *Filicide: The Murder, Humiliation, Mutilation, Denigration, and Abandonment of Children by Parents*, trans. S.H. Rogers, Northvale, NJ: Jason Aronson.

Reed, G.F. (1979) 'Sensory deprivation', in G. Underwood and R. Stevens (eds), *Aspects of Consciousness*, vol. 1, London: Academic Press.

Rosenbluth, D., Harris, M., Osborne, E.L. and O'Shaughnessy, E. (1969) *Your Baby*, London: Corgi.

Rosenfeld, H. (1980) 'Sull'identificazione proiettiva', *Rivista di Psicoanalisi*, 26: 118–39.

Rubinstein, B.B. (1967) 'Explanation and mere description: a metascientific examination of certain aspects of the psychoanalytic theory of motivation', in R.R. Holt, *Motives and Thought: Psychoanalytic Essays in Honor of David Rapaport*, New York: International Universities Press.

Rubinstein, L.H. (1974) 'Crudeltà mentale e fisica verso i bambini', in R. Rossi and R. Speziale-Bagliacca (eds), *Dalla parte di Freud*, Milan: ETAS Kompass.

Salzberger-Wittenberg, I. (1970) *Psycho-Analytic Insight and Relationships: A Kleinian Approach*, London: Routledge and Kegan Paul.

Schafer, R. (1975) 'Psychoanalysis without psychodynamics', *International Journal of Psychoanalysis*, 56: 41–55.

Schrier, A.M., Harlow, M.F. and Stollnitz, F. (eds) (1965) *Behavior of Nonhuman Primates*, New York: Academic Press.

Schultz, D.P. (1965) *Sensory Restriction*, New York: Academic Press.

Schwartz, M. and Day, R.H. (1979) *Visual Shape Perception in Early Infancy*, Chicago: University of Chicago Press.

Searles, H.F. (1965) *Collected Papers on Schizophrenia and Related Subjects*, London: Hogarth.

Segal, H. (1957) 'Notes on symbol formation', *International Journal of Psychoanalysis*, 38: 391–7.

—— (1964a) 'Symposium on fantasy – fantasy and other mental processes', *International Journal of Psychoanalysis*, 45: 191–4.

—— (1964b) *Introduction to the Work of Melanie Klein*, London: Hogarth, 1973.

—— (1978) 'On symbolism', *International Journal of Psychoanalysis*, 59: 315–19.

—— (1979) *Klein*, London: Fontana.

Spitz, R. (1957) *No and Yes: On the Genesis of Human Communication*, New York: International Universities Press.

—— (1965) *The First Year of Life; A Psychoanalytic Study of Normal and Deviant Development of Object Relations*, New York: International Universities Press.

Stern, D. (1985) *The Interpersonal World of the Infant*, New York: Basic Books.

Tauber, E.S. and Green, M.R. (1959) *Prelogical Experience*, New York: Basic Books.

Tustin, F. (1972) *Autism and Childhood Psychosis*, London: Hogarth.
Upson, P. (1975) 'Adolescents and groups, subcultures, countercultures', in S. Meyerson (ed.), *Adolescence: The Crises of Adjustment*, London: Allen and Unwin.
Vincenzi, S. and Imbasciati, A. (1993) 'Contributo alla costruzione di un differenziale semantico per la percezione olfattiva', *Archivio di Psicologia, Neurologia e Psichiatria*, 54: 113–29.
Vurpillot, E. (1972) *The Visual World of the Child*, London: Allen and Unwin, 1976.
Wallerstein, R.S. (1988) 'One psychoanalysis or many?', *International Journal of Psychoanalysis*, 69: 5–21.
—— (1990) 'Psychoanalysis: the common ground', *International Journal of Psychoanalysis*, 71: 3–20.
Warr, P.B. and Knapper, C. (1968) *The Perception of People and Events*, London: Wiley.
Watzlawick, P. (1984) *The Invented Reality: How Do We Know What We Think We Know?*, New York: Norton.
Weich, M.J. (1978) 'Transitional language', in S.A. Grolnick and L. Barkin (eds), *Between Reality and Fantasy*, New York: Jason Aronson.
Winnicott, D.W. (1957) *The Child and the Family: First Relationships*, London: Tavistock.
—— (1958) *Through Paediatrics to Psycho-Analysis*, London: Tavistock.
—— (1965a) *The Family and Individual Development*, London: Tavistock.
—— (1965b) *The Maturational Processes and the Facilitating Environment*, London: Hogarth.
—— (1971) *Playing and Reality*, London: Tavistock.
Zapparoli, G.C. (1979) *La paura e la noia*, Milan: Il Saggiatore.

Index

This index includes a number of 'entries' chosen to illustrate key concepts occurring in the book; some of these are current in contemporary psychoanalysis, while others specifically pertain to the author's theory. Many of these 'entries' have 'sub-entries' that either refer to more limited concepts included under the main concept or indicate connections between the main and other concepts. In other words, the index has been compiled on the basis not only of keywords but also concepts – partly, and in particular, where these are expressed in different words or by circumlocutions. The pages specified may therefore not expressly include the actual keyword, but instead a discussion of the subject in different language. The aim is to provide the reader with an *instrument* to facilitate understanding of the connections between terms and of their interconnections within the theory, as well as to assist with the recapitulation of material by enabling the reader to reread the pages to wich each entry refers. 'Interwoven' with the subject entries are names of authors mentioned in the text. This will help the reader locate precise references to terms and concepts specifically associated with particular authors. It is therefore hoped that this index will prove to be genuinely useful.